INTERNET INVESTIGATIONS
In
ELECTRONICS

by

Cynthia B. Leshin

Prentice Hall
Upper Saddle River, New Jersey   Columbus, Ohio

Editor: Charles E. Stewart, Jr.
Production Editor: JoEllen Gohr
Cover Designer: Julia Zonneveld Van Hook
Production Manager: Patricia A. Tonneman
Marketing Manager: Debbie Yarnell

This book was printed and bound by Quebecor Printing/Book Press. The cover was printed by Phoenix Color Corp.

© 1997 by Prentice-Hall, Inc.
Simon & Schuster/A Viacom Company
Upper Saddle River, New Jersey 07458

Netscape Communication Corporation™ Copyright

Netscape Communication Corporation (NCC), Netscape, Netscape Navigator, and Netsite are trademarks or registered trademarks of Netscape Communications Corporation. Netscape Communication Corporation has granted permission to use screen captures from their home page and to describe Netscape Navigator and its interface. Whereas Netscape's tutorial has been referred to, all efforts have been made not to copy this document.

Printed in the United States of America
10 9 8 7 6 5 4 3 2 1

ISBN: 0-13-496076-9

Prentice-Hall International (UK) Limited, *London*
Prentice-Hall of Australia Pty. Limited, *Sydney*
Prentice-Hall Canada Inc., *Toronto*
Prentice-Hall Hispanoamericana, S. A., *Mexico*
Prentice-Hall of India Private Limited, *New Delhi*
Prentice-Hall of Japan, Inc., *Tokyo*
Simon & Schuster Asia Pte. Ltd., *Singapore*
Editora Prentice-Hall do Brasil, Ltda., *Rio de Janeiro*

DISCLAIMER

While a great deal of care has been taken to provide accurate and current information, the Internet is a dynamic and rapidly changing environment. Information may be in one place today and either gone or in a new location tomorrow. New sites come up daily; others disappear. Some sites provide forwarding address information; others will not. The publisher and author assume no responsibility for errors or omissions. Neither is any liability assumed for damages resulting from the use of this information.

As you travel the information superhighway and find that a resource you are looking for can no longer be found at a given Internet address, there are several steps you can take:

1. Check for a new Internet address or link, often provided on the site of the old address.

2. Use one of the search engines described in Chapter 5 with the title of the Internet resource as keywords.

3. Explore Internet databases such as Yahoo, Magellan, Infoseek, Galaxy or the World Wide Web Virtual Library, which have large directories of Internet resources on Web sites.

The author welcomes readers' feedback, correction of inaccuracies, and suggestions for improvements in subsequent editions. Cynthia Leshin can be contacted by e-mail at: **cleshin@xplora.com**

About the Author

Cynthia Leshin is an educational technologies specialist with her doctorate in educational technology from Arizona State University. Dr. Leshin has her own publishing, training, and consulting company. She has authored three books: *Internet Adventures — Step-By-Step Guide To Finding And Using Educational Resources*, *Netscape Adventures — Step-By-Step Guide To Netscape Navigator and the World Wide Web*, and *Instructional Design: Strategies and Tactics*. The last of these is being used in graduate programs. Her company, XPLORA, publishes the *Internet Adventures* quarterly newsletter to assist teachers with integrating the Internet into the curriculum. Additionally, she is currently writing discipline specific Internet books and Internet-based learning activities for Prentice Hall.

Dr. Leshin has taught computer literacy and Internet classes at Arizona State University West and Estrella Mountain Community College. She currently teaches Internet classes using distance learning technology for Educational Management Group, a Simon & Schuster company. The Internet serves as a tool for teaching and communicating with her students. Her World Wide Web site is a learning resource for students and is also used when making presentations.

Dr. Leshin consults with schools and businesses interested in connecting to the Internet. Her expertise in educational psychology and theories of learning provides her with a unique background for translating complicated technical information into an easy-to-use, easy-to-understand, practical learning resource.

Preface................................

Internet Investigations in Electronics meets the needs of professors, students, and others interested in learning how to use the Internet in the field of Electronics Technology. This cutting edge guide provides step-by-step, easy-to-follow practical information to help you begin using the Internet for finding valuable information.

In this guide you will learn how to easily travel along the information superhighway. As you travel, you will learn how to use

- two Internet navigational tools: Netscape Navigator 2.0 and Microsoft's Internet Explorer.

- the Internet for communicating with others.

- search tools for finding information and locating electronic resources.

- the Internet for career planning.

- the Internet for improving your job opportunities.

Chapter 10 (Learning Adventures) provides hands-on activities for applying and using information and Web resources.

The Appendices provide valuable information for connecting to the Internet and finding an Internet provider.

In this guide you will travel to cool electronic and electrical sites in cyberspace where you will find that viewing multimedia resources is as easy as pointing and clicking your mouse. You will learn more about the practical applications of electronics by visiting leading manufacturing companies and electronic Web sites.

You will learn how to showcase your talents and skills and improve your chances for getting a job by creating an electronic résumé. And, most importantly, you will learn how to use the Internet as a valuable and important tool for your personal and professional life.

Your journey will be divided into two parts:

PART I: Understanding the Internet

PART II: The Web and Electronic Technology

Happy Internet Adventures

Acknowledgments...

The author would like to thank several people for making this guide possible:

I am most grateful to Charles Stewart for the opportunity to write this guide.

Art Gaudette has provided the electronics expertise for this book. He assisted with the research for electronic and electrical Web sites and served as my electronics wizard whenever I had a question.

To my copy editor, Norma Nelson, for teaching me so much about book design and for her many important and useful suggestions to improve my writing.

To my electronics experts, Ronald Reis and Jim Antonakos, for their review of the manuscript.

To all those at Prentice Hall who read the manuscript and made valuable and most appreciated suggestions.

To Todd Haughton and Bob McLaughlin for their artistic support.

To Jill Faber for her input and expertise in career planning and job searching.

To Todd Rossell and Carrie Brandon for opening many new doors for me at Prentice Hall.

To JoEllen Gohr, managing editor, and all the other personnel at Prentice Hall who have transformed these words into this guide.

To my husband, Steve, for his continuing support and for helping to make this Internet adventure possible.

CONTENTS

PART I

Understanding the Internet

CHAPTER 1
What Is the Internet?

• •

In this chapter, you will learn

- ➡ what it means to "be on the Internet."
- ➡ the difference between the Internet and the World Wide Web.
- ➡ Internet addressing protocol—the URL.
- ➡ the three standards used by the World Wide Web.

• •

What Is the Internet?

in'ter·net n
1. world's largest information network **2.** global
web of computer networks **3.** inter-network of many
networks all running the TCP/IP protocol
4. powerful communication tool **5.** giant highway
system connecting computers and the regional
and local networks that connect these computers
syn **information superhighway, infobahn,
data highway, electronic highway, Net,
cyberspace**

The term most frequently used to refer to the Internet is "information superhighway." This superhighway is a vast network of computers connecting people and resources around the world. The Internet is accessible to anyone with a computer and a modem.

The Internet began in 1969 when a collection of computer networks was developed. The first network was sponsored by the United States Department of Defense in response to a need for military institutions and universities to share their research. In the 1970s, government and university networks continued to develop as many organizations and companies began to build private computer networks. In the late 1980s, the National Science Foundation (NSF) created five supercomputer centers at major universities. This special network is the foundation of the Internet today.

Computer networks were initially established to share information among institutions that were physically separate. Throughout the years these networks have grown and the volume and type of information made available to people outside these institutions has also continued to evolve and grow. Today we can exchange electronic mail, conduct research, and look at and obtain files that contain text information, graphics, sound, and video. As more and more schools, universities, organizations, and institutions develop new resources, they are made available to us through our computer networks. These networks make it possible for us to be globally interconnected with each other and to this wealth of information.

What Does It Mean To "Be on the Internet"?

"Being on the Internet" means having full access to all Internet services. Any commercial service or institution that has full Internet access provides the following:

- Electronic mail (e-mail)
- Telnet
- File Transfer Protocol (FTP)
- World Wide Web

Electronic Mail

Electronic mail is the most basic, the easiest to use, and for many people, the most useful Internet service. E-mail services allow you to send, forward, and receive messages from people all over the world, usually at no charge. You can then easily reply to messages, save, file, and categorize received messages.

Electronic mail also makes it possible to participate in electronic conferences and discussions. You can use e-mail to request information from individuals, universities, and institutions.

Telnet
Telnet provides the capability to login to a remote computer and to work interactively with it. When you run a Telnet session, your computer is remotely connected to a computer at another location, but will act as if it were directly connected to that computer.

File Transfer Protocol (FTP)
File Transfer Protocol is a method that allows you to move files and data from one computer to another. File Transfer Protocol, most commonly referred to as FTP, enables you to download magazines, books, documents, free software, music, graphics, and much more.

World Wide Web
The World Wide Web is a collection of standards and protocols used to access information available on the Internet. World Wide Web users can easily access text documents, images, video, and sound.

The Web and the Internet
The World Wide Web (WWW or Web) is a collection of documents linked together in what is called a *hypermedia system*. Links point to any location on the Internet that can contain information in the form of text, graphics, video, or sound files.

Using the World Wide Web requires "browsers" to view Web documents and navigate through the intricate link structure. Currently there are between 30-40 different Web browsers. In this guide you will learn how to use two of the premiere Web browsers—Netscape Navigator and Microsoft's Explorer. Both of these browsers combine a point and click interface design with an "open" architecture that is capable of integrating other Internet tools such as electronic mail, FTP, Gopher, WAIS, and Usenet newsgroups. This architecture makes it relatively easy to incorporate images, video, and sound into text documents.

The World Wide Web was developed at the European Particle Physics Laboratory (CERN) in Geneva, Switzerland. Originally it was developed as a means for physicists to share papers and data easily. Today it has evolved into a sophisticated technology that links hypertext and hypermedia documents.

The Web and the Internet are not synonymous. The World Wide Web is a collection of standards and protocols used to access information available on the Internet. The Internet is the network used to transport information.

The Web uses three standards:

- URLs (Uniform Resource Locators)
- HTTP (Hypertext Transfer Protocol)
- HTML (Hypertext Markup Language)

These standards provide a mechanism for WWW servers and clients to locate and display information available through other protocols such as Gopher, FTP, and Telnet.

URLs (Uniform Resource Locators)

URLs are a standard format for identifying locations on the Internet. They also allow an addressing system for other Internet protocols such as access to Gopher menus, FTP file retrieval, and Usenet newsgroups. URLs specify three types of information needed to retrieve a document:

- the protocol to be used;
- the server address to which to connect; and
- the path to the information.

The format for a URL is: **protocol//server-name/path**

FIGURE 1.1

Sample URLs

World Wide Web URL:	http://home.netscape.com/home/welcome.html
Document from a secure server:	https://netscape.com
Gopher URL:	gopher://umslvma.umsl.edu/Library
FTP URL:	ftp://nic.umass.edu
Telnet URL:	telnet://geophys.washington.edu
Usenet URL:	news:rec.humor.funny

NOTE

The URL for newsgroups omits the two slashes. The two slashes designate the beginning of a server name. Since you are using your Internet provider's local news server, you do not need to designate a news server by adding the slashes.

URL TIPS..

- Do not capitalize the protocol string. For example, the HTTP protocol should be **http://** not **HTTP://**. Some browsers such as Netscape correct these errors. Others do not.

- If you have trouble connecting to a Web site, check your URL and be sure you have typed in the address correctly.

- You do not need to add a slash (/) at the end of a URL such as **http://home.netscape.com** because a slash indicates that there is another path to follow.

HTTP (Hypertext Transfer Protocol)
HTTP is a protocol used to transfer information within the World Wide Web. Web site URLs begin with the http protocol:

http://

This Web URL connects you to Netscape's Home Page.

http://home.netscape.com

HTML (Hypertext Markup Language)
HTML is the programming language used to create a Web page. It formats the text of the document, describes its structure, and specifies links to other documents. HTML also includes programming to access and display different media such as images, video, and sound.

The Adventure Begins...
Now that you have a basic understanding of the Internet you are ready to begin your adventure. Before you can travel and explore the information superhighway you will first need the following:

- an Internet account

- a username and password (required to log onto your Internet account)

- instructions from your institution on how to log on and log off

Getting Started
1. Turn on your computer.
2. Log onto your network using your institution's login procedures.
3. Open Netscape Navigator, Explorer, or the Internet browser that you will be using.

CHAPTER 2
Guided Tour—Internet Browsers

• •

This chapter provides you with a guided tour of the two most widely used Internet browsers—Netscape Navigator and Microsoft's Internet Explorer. You will learn

➥ how to navigate the Internet by using toolbar buttons and pull-down menus.

➥ how to save your favorite Internet sites (URLs) as bookmarks.

• •

Netscape Navigator

Netscape Navigator is a user-friendly graphical browser for the Internet. Netscape makes it possible to view and interact with multimedia resources (text, images, video, and sound) by pointing-and-clicking your mouse on pull-down menus and toolbar buttons.

Netscape Navigator (Version1.0) was developed in 1994 by Marc Andreeseen and others who developed the first graphical Internet browser, Mosaic, at the National Center for Supercomputing Applications (NCSA) at the University of Illinois at Urbana-Champaign. It quickly became the standard and was the premiere Internet information browser in 1995. Netscape Netscape 2.0 was introduced in February 1996 and remains at the head of its field.

Features and Capabilities
Netscape Navigator features include the ability to

• use Netscape as your electronic mail program.
• connect to Gopher, FTP, and Telnet sites without using any additional software.

- read Usenet newsgroups.
- save your favorite Internet addresses (URLs) as bookmarks.
- download images, video, and sound files to your computer desktop.
- view, save, or print the HTML programming code for Web pages as either text or HTML source code.
- use forms for collecting information.
- use plug-in programs, such as JAVA, that extend the capabilities of Netscape.

The Netscape Window (page)

The World Wide Web is unique in that its architecture allows for multimedia resources to be incorporated into a hypertext file or document called a *page*. A Web page or *window* may contain text, images, movies, and sound. Each multimedia resource on a page has associated locational information to link you to the resource. This locational information is called the URL.

The Netscape Navigator 2.0 window includes the following features to assist you with your Internet travels:

- The *Window Title Bar* shows the name of the current document.

- *Page display* shows the content of the Netscape window. A page includes text and links to images, video, and sound files. Links include highlighted words (colored and/or underlined) or icons. Click on a highlighted word or icon to bring another page of related information into view.

- *Frames* is a segmented portion of a Netscape page that contains its own page.

- *Progress Bar* shows the completed percentage of your document layout as your page downloads.

- *Mail Icon* (the small envelope in the bottom-right corner of the Netscape page, or the Mail and News pages) provides you with information on the status of your mail. A question mark next to the mail envelope indicates that Netscape cannot automatically check the mail server for new e-mail messages.

- *Address location* field shows the URL address of the current document.

- *Toolbar* buttons activate Netscape features and navigational aids.

- *Directory* buttons display resources for helping you to browse the Internet.

- Security indicators (*doorkey icon* in the lower-left corner of the window) indicate whether a document is secure (doorkey icon is blue) or insecure (doorkey icon is grey).

The Home Page

The Home Page as shown in Figure 2.1 is the starting point for your journey using a Web browser such as Netscape Navigator. Home pages are created by Internet providers, colleges and universities, schools, businesses, individuals, or anyone who has information they want to make available on the Internet. For example, a college or university may have links to information on the college and courses taught.

FIGURE 2.1
A Home Page for Intel
http://www.intel.com

Navigating With Netscape

This guided tour introduces you to Netscape's graphical interface navigational tools:

- hyperlinks
- toolbar buttons
- pull-down menus

Hyperlinks

When you begin Netscape you will start with a Home Page. Click on highlighted words (colored and/or underlined) to bring another page of related information to your screen.

Images will automatically load onto this page unless you have turned off the **Auto Load Images** found under the **Options** menu. If you have turned off this option you will see this icon that represents an image that can be downloaded.

If you want to view this image, click on this highlighted icon or on the **Images** button.

As you travel the World Wide Web, you will find other icons to represent movies, video, and sound. Click on these icons to download (link) you to these resources.

Toolbar Buttons

Netscape toolbar buttons

🔲 **Back**: Point and click on the **Back** button to go to your previous page.

🔲 **Forward**: This button takes you to the next page of your history list. The history list keeps track of the pages you link to.

Home: This button takes you back to the first opening page that you started with.

Reload: Click on this button to reload the same page that you are viewing. Changes made in the source page will be shown in this new page.

Images: Clicking on this button downloads images onto your current page. Netscape provides you with an option to not download images when you access a page. This makes page downloading faster. If you have selected this option (found in **Options** menu—**Auto Load Images**) and decide that you would like to view an image, just click on the **Images** button.

Open: Use this button to access a dialog box for typing in URLs for Web sites, newsgroups, Gopher, FTP, and Telnet.

Print: Select this button to print the current page you are viewing.

Find: If you are searching for a word in the current page you are viewing, click on the **Find** button for a dialog box to enter the word or phrase.

Stop: This button stops the downloading of Web pages: text, images, video, or sound.

Netscape navigational buttons for exploring the Net

| What's New? | What's Cool? | Handbook | Net Search | Net Directory | Software |

What's New: Visit *What's New* to link to the best new sites.

What's Cool: Netscape's selection of cool Web sites to visit.

Handbook: Links you to on-line Netscape tutorials, references, and index.

Net Search: Clicking on this button links you to available search engines that help find a particular site or document. Search engines use keywords and concepts to help find information in titles or headers of documents, directories, or documents themselves.

Net Directory: Click on this button to explore Internet resources categorized by topic. Some directories cover the entire Internet; some present only what they feel is relevant; others focus on a particular field.

Software: This button connects you to information about Netscape Navigator software: subscription programs, upgrade information, and registration.

Pull-down Menus

Nine pull-down menus offer navigational tools for your Netscape journeys: File, Edit, View, Go, Bookmarks, Options, Directory, Window, and Help (Windows only).

File Menu

Many of the **File** menu options work the same as they do in other applications. You also have options to open a new Netscape window, Home Page, or Internet site.

FIGURE 2.2

Netscape **File** pull-down menu

New Web Browser: Creates a new Netscape window. This window displays the first page you viewed when you connected to Netscape.

New Mail Message: Opens an e-mail composition box that allows you to create and send a message or attach a document to your mail message.

Mail Document (or Mail Frame): Lets you send an e-mail message with the Web page you are viewing attached. The page's URL will be included.

Open Location: Works the same as the **Open** toolbar button. Enter a URL address in the dialog box.

Open File: Provides a dialog box for you to use to open a file on your computer's hard drive. For example, you can open a Web image downloaded to your hard drive without being connected to the Internet.

Close: Closes the current Netscape page. On Windows, this option exits the Netscape application when you close the last page.

Save as... (or Save Frame as): Creates a file to save the contents of the current Internet page you are viewing in the Netscape window. The page can be saved as plain text or in source (HTML) format.

Upload File: Click on this option to upload a file to the FTP server indicated by the current URL. You can also upload images by dragging and dropping files from the desktop to the Netscape window. **NOTE:** This command is active only when you are connected to an FTP server.

Page Setup: Click on this to specify your printing options.

Print: Click on this button to print the current page or frame. To print a single frame, click in the desired frame.

Print Preview (Windows only): Previews the printed page on the screen.

Exit (on Macintosh—Quit): Exits the Netscape application.

Edit Menu

The **Edit** menu makes it possible to cut and paste text from a Web page to your computer's clipboard. This option can be used to copy and paste text from a page or frame to a word processing document or another application of your choice. The options under this menu are similar to what you have available to you in many of your computer software applications under their **File** menu (i.e., word processing, desktop publishing, and graphics applications).

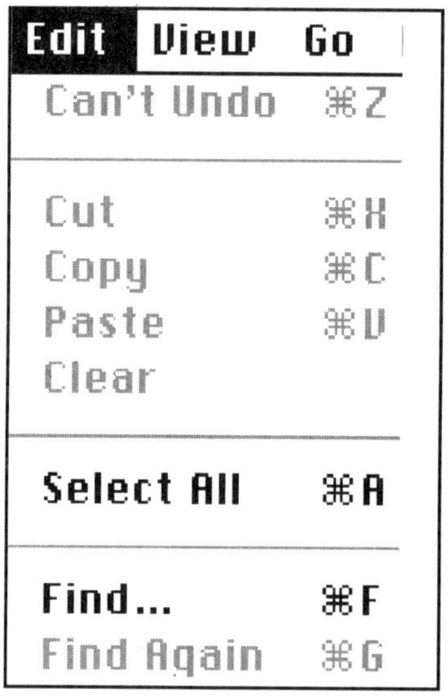

FIGURE 2.3
Netscape **Edit** menu

Edit	View	Go
Can't Undo		⌘Z
Cut		⌘X
Copy		⌘C
Paste		⌘V
Clear		
Select All		⌘A
Find...		⌘F
Find Again		⌘G

- **Undo..** (or **Can't Undo**): May reverse the last action you performed.

- **Cut:** Removes what you have selected and places it on the clipboard.

- **Copy:** Copies the current selection to the computer's clipboard.

- **Paste:** Puts the current clipboard's contents in the document you are working on.

- **Clear** (for the Macintosh only): Removes the current selection.

Select All: Selects all you have indicated by using the application's selection markers. May be used to select items before you cut, copy, or paste.

Find: Lets you search for a word or phrase within the current Web page.

Find Again: Searches for another occurrence of the word or phrase specified when you used the **Find** command.

View Menu

The **View** menu offers options for viewing images, the Netscape page, HTML source code, and information on the current Web's document structure.

FIGURE 2.4
View menu options from Netscape

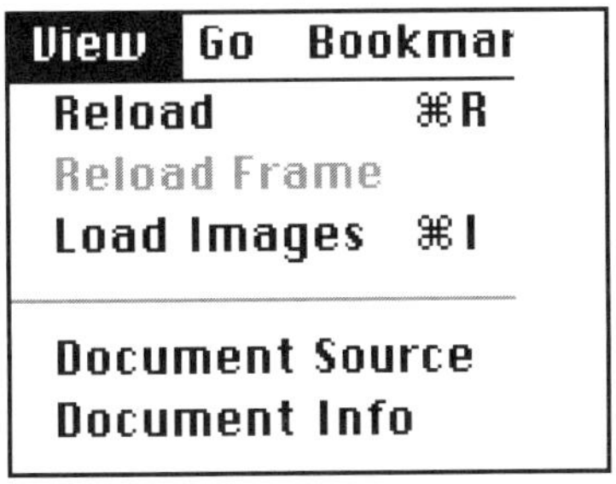

View	Go	Bookmar
Reload		⌘R
Reload Frame		
Load Images		⌘I
Document Source		
Document Info		

Reload: Downloads a new copy of the current Netscape page you are viewing to replace the one originally loaded. Netscape checks the network server to see if any changes have occurred to the page.

Reload Frame: Downloads a new copy of the currently selected page within a single frame on a Netscape page.

Load Images: If you have set **Auto Load Images** in your Netscape **Options** menu, images from a Web page will be automatically loaded. If this option has not been selected, choose **Load Images** to display the current Netscape page.

Refresh (Windows only): Downloads a new copy of the current Netscape page from local memory to replace the one originally loaded.

17

■ **Document Source:** Selecting this option provides you with the format of HTML (HyperText Markup Language). The HTML source text contains programming commands used to create the page.

■ **Document Info:** Produces a page in a separate Netscape window with information on the current Web document's structure and composition, including title, location (URL), date of the last modification, character set encoding, and security status.

Go Menu

The **Go** menu has Netscape navigational aids.

FIGURE 2.5
Netscape **Go** menu

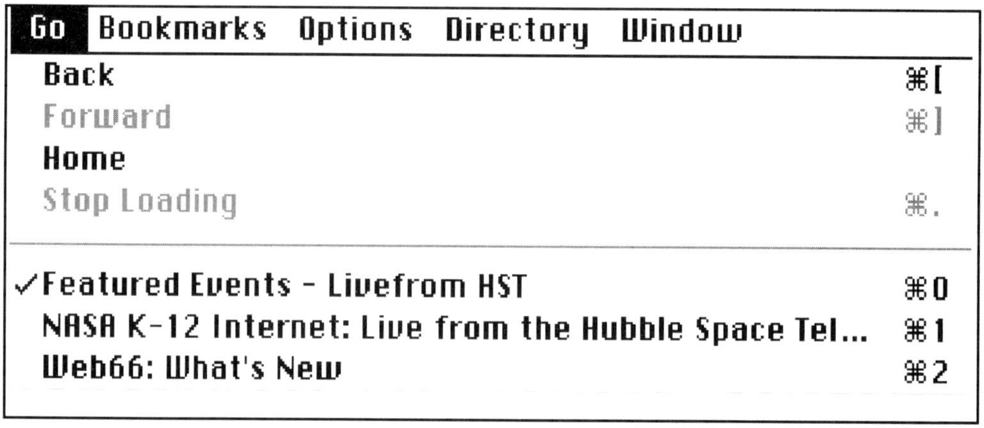

Go	Bookmarks	Options	Directory	Window	
Back					⌘[
Forward					⌘]
Home					
Stop Loading					⌘.
✓Featured Events – Livefrom HST					⌘0
NASA K-12 Internet: Live from the Hubble Space Tel...					⌘1
Web66: What's New					⌘2

■ **Back**: Takes you back to the previous page in your history list. Same as the **Back** button on the toolbar. The history list keeps track of all the pages you link to.

■ **Forward**: Takes you to the next page of your history list. Same as the **Forward** button on the toolbar.

■ **Home**: Takes you to the Home Page. Same as the **Home** button on the toolbar.

Stop Loading: Stops downloading the current page. Same as the **Stop** button.

History Items: A list of the titles of the places you have visited. Select menu items to display their page. To view the History list, select the **Window** menu and then choose **History.**

Bookmarks Menu

Bookmarks makes it possible to save and organize your favorite Internet visits. Opening this pull-down menu allows you to view and download your favorite pages quickly.

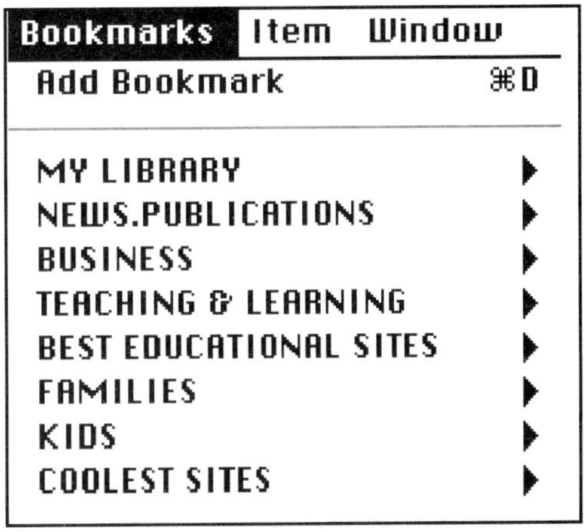

FIGURE 2.6
Netscape **Bookmark** menu

Bookmarks Item Window	
Add Bookmark	⌘D
MY LIBRARY	▶
NEWS.PUBLICATIONS	▶
BUSINESS	▶
TEACHING & LEARNING	▶
BEST EDUCATIONAL SITES	▶
FAMILIES	▶
KIDS	▶
COOLEST SITES	▶

Add Bookmark: Click on **Add Bookmark** to save this page in your bookmark list. Behind the scenes, Netscape saves the URL address so you can access this page by pointing-and-clicking on the item in your list.

Bookmark Items: Below **Add Bookmark**, you will see a list of your saved pages. Point and click on any item to bring this page to your screen.

To view your bookmarks, add new bookmark folders, arrange the order of your bookmarks, or to do any editing, select the **Window** menu and choose **Bookmarks**.

Options Menu

The **Options** menu offers customization tools to personalize your use of Netscape Navigator. Several uses for these customization tools include:

- showing the toolbar buttons.
- showing the URL location of a page.
- showing the Directory buttons.
- automatic loading of images.
- selecting styles for pages to appear.
- selecting which Home Page you want to appear when you log onto Netscape.
- selecting link styles (colors).
- selecting your news server to interact with Usenet newsgroups.
- setting up e-mail on Netscape.

There are additional customization tools available that are more advanced. Refer to the Netscape on-line handbook for more information on **Options** and **Preferences**.

> ### NOTE
> Before you can use the e-mail and Usenet newsgroup tools available in Netscape, you will need to customize the **Mail and News Preferences**.

FIGURE 2.7
Netscape **Options** menu

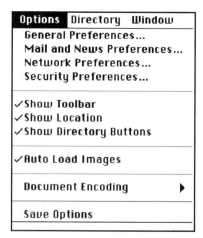

General Preferences: Presents tab buttons for selecting preferences. Each tab presents a panel for customizing Netscape's operations for your personal needs, preferences, and interests.

Mail and News Preferences: Panel for entering information on your mail and news server, so you can use Netscape to send and receive e-mail and to participate in Usenet newsgroups.

Network Preferences: Options for cache, network connections, and proxy configurations.

Security Preferences: Panel for setting security features.

Show Toolbar: If selected, the Toolbar buttons are visible on the Netscape page.

Show Location: If selected, the URL location for the page is displayed.

Show Directory Buttons: If selected, the Directory buttons are visible.

Show Java Console (Windows only): If selected, displays the Java Console window.

Auto Load Images: If selected, images embedded into a page will be loaded automatically. If not checked, images can be loaded by clicking on the **Load Images** button. Deselecting this option increases the speed of downloading a page.

Document Encoding: Lets you select which character set encoding a document uses when document encoding is either not specified or unavailable. The proportional and fixed fonts are selected using the **General Preferences/Fonts** panel.

Save Options: Click on this option to save the changes you made to any of the above options.

Directory Menu

The **Directory** pull-down menu directs you to a few navigational aids to help you begin your Web exploration.

FIGURE 2.8
Netscape **Directory** menu

Directory	Window
Netscape's Home	
What's New?	
What's Cool?	
Netscape Galleria	
Internet Directory	
Internet Search	
Internet White Pages	
About the Internet	

Netscape's Home: Takes you to the Netscape Home Page.

What's New: Click on this item to see what's new on the Internet.

What's Cool: Netscape's selection of interesting places to visit.

Netscape Galleria: A showcase of Netscape customers who have built Net sites using Netscape Server software. Visit the Galleria to learn more about how to build and maintain innovative Web sites.

Internet Directory: Same as the Internet Directory button. Links you to Internet directories for finding information and resources.

Internet Search: Connects you to many of the best on-line search engines.

Internet White Pages: Links you to tools to help you find people connected to the Internet.

About the Internet: Links to resources to help you learn more about the Internet.

Window Menu

The **Window** menu makes it possible for you to navigate easily between your e-mail, Usenet news, and Bookmarks windows, and to see and visit places you have already traveled.

FIGURE 2.9
Netscape **Window** menu

| Macintosh **Window** | Windows **Window** |

 **Netscape Mail:** Click on this option to access the Netscape e-mail program.

 **Netscape News:** Click on this option to access the Usenet newsgroups.

 **Address Book:** Displays an Address Book window for use with the e-mail program.

 **Bookmarks:** Displays bookmarks and pull-down menus for working with or editing your bookmarks.

 **History:** Displays a history list of the pages (their titles and URLs) that you have recently viewed. Select an item and press the **Go To** button (or double-click) to revisit the page.

Internet Explorer

Now that you're familiar with Internet navigation using Netscape, you will be able to transfer that knowledge to the use of other Internet browsers. Most browsers have similar or the same navigational tools in the form of toolbar buttons and pull-down menus. Microsoft's Internet Explorer is another widely used and highly sophisticated browser that is integrated with the Windows 95 operating environment. Explorer is the primary Internet browser for America On-line (AOL) and CompuServe. Notice in Figure 2.10 how similar the navigational tools are to those of Netscape's Navigator.

FIGURE 2.10
Microsoft's Internet Explorer window

Navigating With Internet Explorer

Toolbar buttons and pull-down menus are your Internet navigational tools when using Internet Explorer.

Toolbar Buttons

 Open: Accesses a dialog box for typing in URLs, documents, or folders for Windows to open.

 Print: Prints the page you are viewing.

 Send: Information services for using Microsoft's fax, e-mail, Netscape Internet transport, or Microsoft's Network On-line Services.

 Back/Forward: Takes you either back to your previous page or forward to the next page in your history list.

 Stop: Stops the downloading of a Web page: text, images, video, or sound.

 Refresh: Brings a new copy of the current Explorer page from local memory to replace the one originally loaded.

 Open Start Page: Takes you back to the first opening page.

 Search the Internet: Click this button for a list of search services to help you find information on the Internet.

 Read Newsgroups: This option brings up a list of Usenet newsgroups available from your Internet provider or college/university.

Open Favorites: Click this button to see a list of your favorite URLs.

 Add to Favorites: Click on this button to add a favorite URL to your list.

 Use Larger/Smaller Font: Increase or decrease the size of the font on the page you are viewing.

 Cut: Removes what you have selected and places it on the clipboard.

 Copy: Copies the current selection to the computer's clipboard.

 Paste: Puts the current clipboard's contents in the document you are working on.

Pull-down Menus

Pull-down menus offer navigational tools for your Internet exploration. Some of the options are similar to the toolbar buttons: File, Edit, View, Go, Favorites, Help.

> **NOTE**
> The pull-down menus will *not* be discussed or shown *unless* their functions differ significantly from the discussion of pull-down menus under Netscape Navigator.

File Menu

Explorer's **File** menu provides options for connecting to new Internet sites, printing Web pages, creating desktop shortcuts to your favorite Web pages, and to finding information about the page you are viewing.

FIGURE 2.11
Explorer's **File** pull-down menu

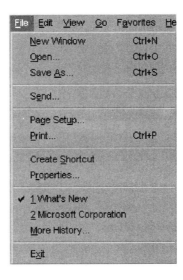

 **Create Shortcut**: Select this option to create a shortcut to the current page that will be placed on your desktop.

 **Properties**: Provides you with general information about the page you are viewing, including security information.

Edit Menu

The **Edit** menu offers cut, copy, and paste options as well as a find command for keywords searches.

FIGURE 2.12
Explorer's **Edit** menu

View Menu

The **View** menu provides options for how your Explorer page appears. **Toolbar**, **Address Bar**, and **Status Bar** provide options for viewing or not viewing these Explorer tools.

FIGURE 2.13
Explorer's **View** menu
with active tools checked

Go Menu

The **Go** menu provides options for moving forward to the next page in your history list or backward to a previous page.

FIGURE 2.14
Explorer's **Go** menu
displaying navigational options

■ **Start Page**: Takes you back to the first opening page you started with.

■ **Search the Internet**: Takes you to search tools for finding information on the Internet.

■ **Read Newsgroups**: This option takes you to Explorer's news reader for Usenet newsgroups.

Favorites Menu

Explorer's **Favorites** list is the same as Netscape's Bookmarks or what other browser's refer to as a hotlist.

FIGURE 2.15
Explorer's **Favorites** menu

■ **Add To Favorites**: Select this option to add the URL of a Web site to Explorer's Favorites list.

■ **Open Favorites:** Use this option to select a URL for Explorer to open.

Help menu

The **Help** menu provides help with using Internet Explorer.

FIGURE 2.16
Explorer's **Help menu**

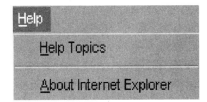

FIGURE 2.17

Explorer's **Help** Contents panel

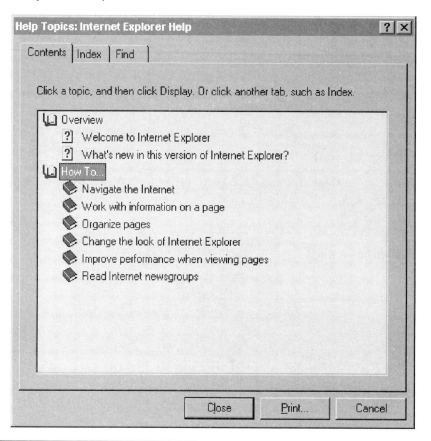

CHAPTER 3
Hands-on Practice

In this chapter, you will practice using Netscape and/or Explorer for

- navigating the Internet.
- organizing and using bookmarks.
- exploring the Internet.

Practice 1: Browsing the Internet

In this guided practice you will use Netscape Navigator or Explorer to

- connect to World Wide Web sites and Home Pages;
- use pull-down menus and navigational toolbar buttons to navigate World Wide Web sites; and
- save bookmarks of your favorite pages.

1. *Log onto your Internet account.* When you have connected, open the Netscape Navigator or Explorer browser by double-clicking on the application icon.

Netscape Navigator Icon

Microsoft Explorer Icon

You will be taken to a Home Page. Notice the Location/Address URLs in Figure 3.1 and Figure 3.2. This Home Page may belong to Netscape Communications Corporation (**http://home.netscape.com**) or Microsoft (**http://www.microsoft.com**), or it may have been designed by your college or university. Look at the top of the Home Page in the Title Bar to see whose Home Page you are visiting.

URLs are a standard for locating Internet documents. Highlighted text on Netscape pages contains built-in URL information for linking to that information. You can also type in new URL text to link a page.

FIGURE 3.1

Netscape Navigator toolbar buttons

FIGURE 3.2

Microsoft's Internet Explorer toolbar buttons

2. *Begin exploring* the World Wide Web by using Netscape's toolbar buttons and pull-down menus. Click on the **What's New** button. You will see a list of highlighted underlined links to Web sites. Click on a link and EXPLORE. HAVE FUN! If you are using Explorer, investigate the Home Page that you are viewing.

3. *Save your favorite pages* by making a bookmark or an addition to your Favorites List.

When you find a page that you may want to visit at a later time, click on the pull-down menu, **Bookmarks**. Next, click on the menu item **Add Bookmark**. (Explorer—select the **Favorites** menu.)

Click on the **Bookmarks** (**Favorites**) pull-down menu again. Notice the name of the page you marked listed below the **View Bookmarks**

menu item. To view this page again, select the **Bookmarks** pull-down menu and click on the name of the page you saved.

4. Continue your exploration by clicking on the **What's Cool** button.

5. After you have linked to several pages, click on the **Go** pull-down menu. Notice the listing of the places you have most recently visited. If you want to revisit any of the pages you have already viewed, click on the name of the Web site.

Practice 2: Organizing and Using Bookmarks

In this practice you will learn how to organize, modify, save, and move bookmark files. If you are using Explorer, save your favorite URLs by using either the **Favorites** button or the **Favorites** menu.

Before you can organize and work with bookmark files, you must access Netscape's **Bookmark** window. There are two ways to access the **Bookmark** window:

- Go to the **Bookmarks** pull-down menu and select **Go To Bookmarks**; or

- Go to the **Window** pull-down menu and select **Bookmarks**.

1. *Organizing your bookmarks.* Before you begin saving bookmarks it is helpful to consider how to *organize* saved bookmarks. Begin by thinking of categories that your bookmarks might be filed under such as Software, Business, Education, Entertainment, Research, etc. For each category make a folder. These are the steps for making your bookmark folders.

a. Go to the **Bookmarks** menu and select **Go To Bookmarks**, or go
to the **Window** menu and select **Bookmarks**.

FIGURE 3.3
The Netscape **Bookmarks**
window

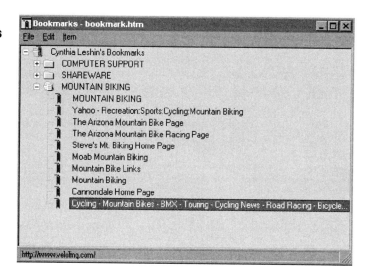

Notice the Web sites saved in the bookmarks folders in Figure 3.3.
This Bookmarks window provides you with three new menus for
working with your bookmarks: **File**, **Edit**, and **Item**.

b. Create a new folder for a bookmark category by selecting the **Item**
menu (Fig. 3.4).

FIGURE 3.4
Opened **Item** menu from within
the Bookmarks window

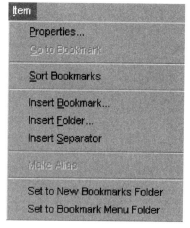

c. Select **Insert Folder** (see Fig. 3.5).

FIGURE 3.5
Insert Folder window

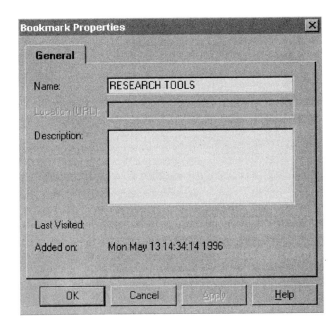

d. Type in the name of your folder in the Name dialog box.

e. Enter in any description of the bookmark folder.

f. Click OK.

2. *Adding bookmarks to a folder*. Netscape provides an option for identifying which folder you would like to select to drop your bookmarks in.

a. Select the folder you would like to add your new bookmarks to by clicking on the name of the folder once. The folder should now be highlighted.

b. Go to the **Item** menu and select **Set to New Bookmarks Folder** shown near the bottom of Figure 3.4.

c. Go back to your Bookmark window and notice how this newly identified folder has been marked with a colored Bookmark identifier. All bookmarks that you add will be placed in this folder until you identify a new folder.

3. *Modifying the name of your bookmark.* Bookmark properties contain the name of the Web site and the URL. You may want to change the name of the bookmark to indicate more clearly the information available at this site. For example, the bookmark name *STCil/HST Public Information* has very little meaning. Changing its name to *Hubble Space Telescope Public Information* is more helpful later when selecting from many bookmarks.

a. To change the name of a bookmark, select the bookmark by clicking on it once.

b. Go to the **Item** menu from within the Bookmark window.

c. Select **Properties**.

FIGURE 3.6
Properties window
from Bookmark **Item**
options

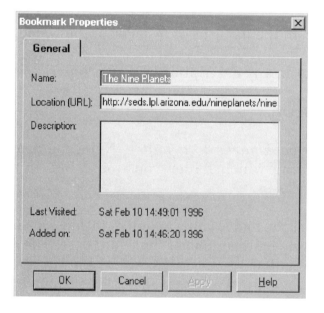

d. Enter in the new name for your bookmark by either deleting the text shown in Figure 3.6 or begin typing the new name when the highlighted text is visible.

e. Notice the URL for the bookmark; you can also enter in a new description for the URL.

4. *Making copies of your bookmarks for adding to other folders.* Occasionally you will want to save a bookmark in several folders. There are two ways to do this:

a. Select the bookmark that you would like to copy. Go to the **Edit** menu from within the Bookmark window and select **Copy**. Select the folder where you would like to place the copy of the bookmark. Go to the **Edit** menu and select **Paste**.

b. Make an alias of your bookmark by selecting **Make Alias** from the **Item** menu. When the alias of your bookmark as been created, move the alias bookmark to the new folder (see "Note").

> ## NOTE
> Bookmarks can be moved from one location to another by dragging an existing Bookmark to a new folder.

5. *Deleting a bookmark.* To remove a bookmark:

a. Select the bookmark to be deleted by clicking on it once.

b. Go to the **Edit** menu from within the Bookmark window.

c. Choose either **Cut** or **Delete**.

6. *Exporting and saving bookmarks.* Netscape provides options for making copies of your bookmarks to either save as a backup on your hard drive, to share with others, or to use on another computer.

Follow these steps for exporting or saving your bookmarks to a floppy disk.

a. Open the **Bookmark** window.

b From within the Bookmark window, go to the **File** menu. Select **Save As**.

c. Designate where you would like to save the bookmark file—on your hard drive or to a floppy disk—in the **Save in** box.

FIGURE 3.7

Netscape Bookmark window for saving bookmark files

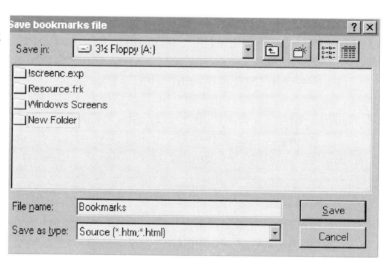

d. Enter in a name for your bookmark file in the **File name** dialog box.

e. Click **Save**.

7. *Importing Bookmarks.* Bookmarks can be imported into Netscape from a previous Netscape session saved on a floppy disk.

a. Insert the floppy disk with the bookmark file into your computer.

b. Open the **Bookmark** window.

c. From within the Bookmark window, go to the **File** menu and select **Import** (see Fig. 3.8).

d. Designate where the bookmark file is located: The **Look in** window displays a floppy disk or you can click on the scroll arrow to bring the hard drive into view.

FIGURE 3.8

Import window allows bookmark files from a floppy disk to be imported into your Netscape application

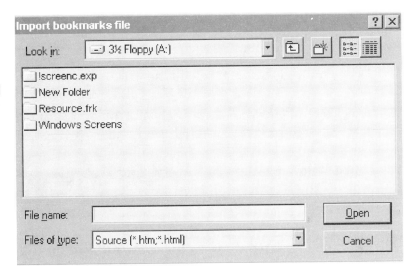

e. *Click on **Open**.* The bookmarks will now be imported into your Netscape bookmark list.

Practice 3:
Exploring the Internet With Your Web Browser

In this practice you will enter in URL addresses to link to World Wide Web (WWW) sites.

There are three options in Netscape for entering in a URL:

- the **Location** text field;
- the **File** menu—**Open Location**; or
- the **Open** toolbar button.

If you are using Explorer, select the **Open** button.

1. Select one of the above options to bring you to the window where you can enter your choice of URL text.

2. Listed below are several interesting Web sites to visit. Type a URL and EXPLORE. Remember to save your favorite sites as Bookmarks/ Favorites.

Awesome List: **http://www.clark.net/pub/journalism/awesome.html**

CityNet: **http://www.city.net**

ESPNetSportsZone: **http://espnet.sportzone.com**

NASA: **http://www.nasa.gov**

Time Warner Pathfinder:
http://www.timeinc.com/pathfinder/Greet.html

The White House: **http://www.whitehouse.gov**

Wired: **http://www.hotwired.com**

CHAPTER 4
Chatting on the Net

• •

Although the Internet was created as a research network, it soon became popular for chatting and discussing work-related topics and hobbies. In this chapter you will learn about how to communicate with others using

- ➥ listserv mailing lists.
- ➥ Usenet newsgroups.
- ➥ Internet Relay Chat (IRC).
- ➥ Internet phones.

• •

People today are on the Internet because they value and enjoy the interactivity and the relationships they build within the virtual community of cyberspace. The way companies, institutions, and individuals communicate has changed. Internet communication involves five major services: electronic mail, electronic discussion groups (listservs and Usenet), Internet Relay Chat (IRC), Internet phones, and desktop Internet videoconferencing. E-mail and electronic discussion groups are delayed response media. IRC, Net phones, and desktop videoconferencing are real-time media. Net phones and videoconferencing are usually used for private conversations and IRC as a public forum. Electronic mail is most often used for private conversations; electronic discussion groups are used for public conversation.

Listserv Mailing Lists

With so much attention on the World Wide Web, many new Internet users miss learning about electronic mailing lists (also referred to as lists, listservs, or discussion groups) as an Internet resource for finding and sharing information. Electronic mailing lists began in the 1960s when scientists and educators used the Internet to share information and research. Early programs, known as *listservs*, ran on mainframe computers and used e-mail to send reports or studies to a large group of users.

Today, listservs perform the same function—the sharing of information. There are hundreds of special interest lists where individuals can join a virtual community to share and discuss topics of mutual interest.

What Is a Listserv Mailing List?

A *listserv* is the automated system that distributes electronic mail. E-mail is used to participate in electronic mailing lists. Listservs perform two functions:

- Distributing text documents stored on them to those who request them, and
- Managing interactive mailing lists.

Listservs and text documents

A listserv can be used to distribute information, in the form of text documents, to others. For example, on-line workshops may make their course materials available through a listserv. The listserv is set up to distribute the materials to participants at designated times. Other examples of documents available throughs a listserv include: a listing of all available electronic mailing lists, Usenet newsgroups, electronic journals, and books.

Interactive mailing lists

Interactive mailing lists provide a forum where individuals who share interests can exchange ideas and information. Any member of the group may participate in the resulting discussion. This is no longer a one-to-one communication like your e-mail, but rather a one-to-many communication.

Electronic mail written in the form of a report, article, abstract, reaction, or comment is received at a central site and then distributed to the members of the list.

How Does a Mailing List Work?

The mailing list is hosted by a college, university, or institution. The hosting institution uses its computer system to manage the mailing list.

Here are a few of the management functions of a listserv:

- receiving requests for subscription to the list;
- placing subscribers' e-mail addresses on the list;
- sending out notification that the name has been added to the list;
- receiving messages from subscribers;
- sending messages to all subscribers;
- keeping a record (archive) of activity of the list; and
- sending out information requested by subscribers to the list.

Mailing lists have administrators that may be either a human or a computer program. One function of the administrator is to handle subscription requests. If the administrator is human, you can join the mailing list by communicating in English via an e-mail message. The administrator in turn has the option of either accepting or rejecting your subscription request. Frequently lists administered by a human are available only to a select group of individuals. For example, an executive board of an organization may restrict its list to its members.

Mailing lists administered by computer programs called listservs usually allow all applicants to subscribe to the list. You must communicate with these computer administrators in listserv commands. For the computer administrator to accept your request, you must use the exact format required. The administrative address and how to subscribe should be included in the information provided about a list.

How To Receive Documents From a Listserv

E-mail is used to request text documents distributed by a listserv. The e-mail is addressed to the listserv *administrative address*. In the body of the message a command is written to request the document. The most common command used to request a document is "send" or "get." The command is then followed by the name of the document that you wish to receive. A command to request a list of interesting mailing lists might look like this:

"get" or "send" <name of document>

or

get new-list TOP TEN

How To Join a Listserv Mailing List

To join an interactive mailing list on a topic of interest, send an e-mail message to the list administrator and ask to join the list. Subscribing to an electronic mailing list is like subscribing to a journal or magazine.

- Mail a message to the journal with a request for subscription.

- Include the address of the journal and the address to which the journal will be mailed.

All electronic mailing lists work in the same way.

- E-mail your request to the list administrator at the address assigned by the hosting organization.

- Place your request to participate in the body of your e-mail where you usually write your messages.

- Your return address will accompany your request in the header of your message.

- Your subscription will be acknowledged by the hosting organization or the moderator.

- You will then receive all discussions distributed by the listserv.

- You can send in your own comments and reactions.

- You can unsubscribe (cancel your subscription).

The command to subscribe to a mailing lists looks like this.

subscribe <name of list> _<your name>_

or

subscribe EDUPAGE Cynthia Leshin

The unsubscribe command is similar to the subscribe command.

unsubscribe <name of list> *<your name>*

Active lists may have 50-100 messages from list participants each day. Less active mailing lists may have several messages per week or per month. If you find that you are receiving too much mail or the discussions on the list do not interest you, you can unsubscribe just as easily as you subscribed. If you are going away, you can send a message to the list to hold your mail until further notice.

> **NOTE**
> In Chapter 6 you will find a list of electronic listservs and Usenet newsgroups. Use the information provided in this chapter to find listservs and newsgroups of interest to you.

Finding Listserv Mailing Lists
World Wide Web Site For Finding Mailing Lists
One of the best resources for helping you to find mailing lists is this World Wide Web site

http://www.tile.net/tile/listserv/index.html

E-Mail A Request For Listservs On A Topic
To request information on listserv mailing lists on a particular topic, send an e-mail message to

LISTSERV@vm1.nodak.edu

In the message body type: **LIST GLOBAL** / *keyword*

To find electronic mailing lists you would enter
LIST GLOBAL/ *electronics*

Usenet Newsgroups

What Are Newsgroups?

In the virtual community of the Internet, Usenet newsgroups are analogous to a cafe where people with similar interests gather from around the world to interact and exchange ideas. Usenet is a very large, distributed bulletin board system (BBS) that consists of several thousand specialized discussion groups. Currently there are over 20,000 newsgroups with 20 to 30 more added weekly.

You can subscribe to a newsgroup, scan through the messages, read messages of interest, organize the messages, and send in your comments or questions—or start a new one.

Usenet groups are organized by subject and divided into major categories.

Category	Topic Area
alt.	no topic is off limits in this alternative group
comp.	computer-related topics
misc.	miscellaneous topics that don't fit into other categories
news.	happenings on the Internet
rec.	recreational activities/hobbies
sci.	scientific research and associated issues
soc.	social issues and world cultures
talk.	discussions and debates on controversial social issues

In addition to these categories there are local newsgroups with prefixes that indicate their topic or locality.

Some newsgroups are moderated and reserved for very specific articles. Articles submitted to these newsgroups are sent to a central site. If the article is approved, it is posted by the moderator. Many newsgroups have no moderator and there is no easy way to determine whether a group is moderated. The only way to tell if a group is moderated is to submit an article. You will be notified if your article has been mailed to the newsgroup moderator.

What Is the Difference Between Listserv Mailing Lists and Usenet Newsgroups?

One analogy for describing the difference between a listserv mailing list and a Usenet newsgroup is to compare the difference between having a few intimate friends over for dinner and conversation (a listserv) vs. going to a Super Bowl party to which the entire world has been invited (newsgroups). A listserv is a smaller, more intimate place to discuss issues of interest. A Usenet newsgroup is much larger and much more open to "everything and anything goes." This is not to say that both do not provide a place for valuable discussion. However, the size of each makes the experience very different.

A listserv mailing list is managed by a single site, such as a university. Subscribers to a mailing list are automatically mailed messages that are sent to the mailing list submission address. A listserv would find it difficult to maintain a list for thousands of people.

Usenet consists of many sites that are set up by local Internet providers. When a message is sent to a Usenet site, a copy of the message that has been received is sent to other neighboring, connected Usenet sites. Each of these sites keeps a copy of the message and then forwards the message to other connected systems. Usenet can therefore handle thousands of subscribers.

One advantage of Usenet groups over a mailing list is that you can quickly read postings to the newsgroup. When you connect to a Usenet newsgroup and see a long list of articles, you can select only those that interest you. Unlike a mailing list, Usenet messages do not accumulate in your mailbox, forcing you to read and delete them. Usenet articles are on your local server and can be read at your convenience.

Netscape and Usenet Newsgroups

Netscape supports Usenet newsgroups. You can subscribe to a newsgroup, read articles posted to a group, and reply to articles. You can determine whether your reply is sent to the individual author of the posted article or to the entire newsgroup.

Netscape has an additional feature. Every news article is scanned for references to other documents called URLs. These URLs are shown as active hypertext links that can be accessed by clicking on the underlined words.

Newsgroups have a URL location. These URLs are similar, but not identical, to other pages. For example, the URL for a recreational backcountry newsgroup is **news:rec.backcountry**. The server protocol is **news:** and the newsgroup is **rec.backcountry.**

Newsgroups present articles along what is called a "thread." The thread packages the article with responses to the article. Each new response is indented one level from the original posting. A response to a response is indented another level. Newsgroups' threads, therefore, appear as an outline.

Buttons on each newsgroup page provide the reader with controls for reading and responding to articles. Netscape buttons vary depending on whether you are viewing a page of newsgroup listings or a newsgroup article.

Netscape News Window for Usenet News

To display the News window, go to the **Window** menu and select **Netscape News**.

FIGURE 4.1

The Netscape **News** window

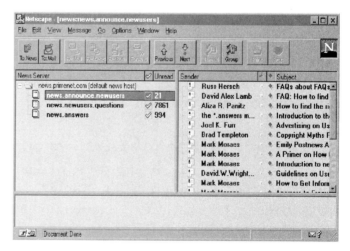

Notice that you have new options in the form of toolbar buttons and pull-down menus for receiving, reading, replying, and sending messages to newsgroups. Netscape News works in much the same way as Netscape Mail.

Netscape News window buttons

To: News: Displays a Message Composition window for creating a new message posting for a newsgroup.

To: Mail: Displays a Message Composition window for creating a new mail message.

Re: Mail: Click on this button to reply to the current newsgroup message (thread) you are reading.

Re: Both: Displays a Message Composition window for posting a reply to the current message thread for the entire newsgroup and to the sender of the news message.

Forward: Displays the Message Composition window for forwarding the current news message as an attachment. Enter the e-mail address in the **Mail To** field.

Previous: Brings the previous unread message in the thread to your screen.

Next: Brings the next unread message in the thread to your screen.

Thread: Marks the message threads you have read.

Group: Marks all messages read.

Print: Prints the message you are reading.

Stop: Stops the current transmission of messages from your news server.

Netscape News Menus

When you select Netscape News you will receive not only new toolbar buttons but also different pull-down menus for interacting with the Netscape news reader: File, Edit, View, Message, Go, Options, Window, and Help.

> **NOTE**
>
> This guide does not provide detailed information on using Netscape Navigator for Usenet newsgroups. For more information refer to *Netscape Adventures—Step-By-Step Guide To Netscape Navigator And The World Wide Web* by Cynthia Leshin.
>
> For information on using Internet Explorer for newsgroups, see Explorer's on-line **Help** menu.

Reading Usenet News With Netscape Navigator

Netscape Navigator 2.0 provides four easy-to-access newsgroups.

- If you know the name of the newsgroup, type the URL in the location field of the Netscape main menu.

- From within the Netscape News window, go to the **File** menu and select **Add Newsgroup**. Enter the name of the newsgroup in the dialog box.

- From within the Netscape News window, go to the **Options** menu and select **Show All Newsgroups**. From this list, select a newsgroup and check the **Subscribe** box beside the newsgroup name.

- From a World Wide Web site, click on a link to a newsgroup or a newsgroup message.

FIGURE 4.2
Netscape News window
Options menu

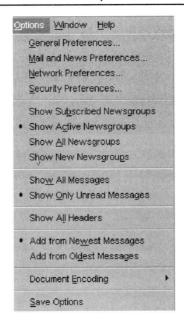

Chats

Chats are programs that allow you to talk to many people at the same time from all over the world. Internet Relay Chat (IRC) is the most widely used program and has become one of the most popular Internet services. IRC has produced a new type of virtual community formed mainly by many young people. IRC has in fact produced its own subculture of addicted chat junkies who spend most of their time on-line chatting. Although information may be exchanged on any topic and users can send and receive files, the primary use of IRC seems to be idle chatter and gossip.

Many Internet access providers make IRC available to subscribers. Some institutions have IRC client programs installed. To connect to IRC, users merely type **irc**. If your institution does not have an IRC client program, you can Telnet to a public IRC server and chat from there. Some World Wide Web sites will have chat rooms for interactive discussion on topics of interest. For example, Time Warner's Pathfinder Web site has a chat room for discussing news of the day. Wired magazine has a chat room open for discussion.

After you have connected to an IRC, you will have to choose an on-line name known as a *nickname* to identify yourself. You will be known by your nickname. Next, you select a group or discussion topic, known as a *channel*, to join. There are any number of channels in IRC and any number of people within a channel. Some channels exist all the time; others come and go.

Conversations within chats are text-based. Users type in their message line by line. As a line is being typed others on the channel see the message. Messages cannot be edited before they are sent to others on the channel. Anyone on the channel can respond to a message as it is revealed on their computer screen by merely typing in their response line by line.

Commercial services such as America Online, CompuServe, Prodigy, and Microsoft Network offer chat rooms for their members to communicate and meet others who have similar interests. Chat rooms with these services can be public or private. Public rooms are created by the service provider and tend to have focused discussion topics. Some of these rooms are hosted, others are not. Some of these chat rooms are available on a regular basis, others are created for special events such as a guest who is on-line for a forum for several hours.

Private chat rooms are created by members and can hold between 2 and 25 or more registered on-line users. Private chat rooms may be used for a meeting or just a casual chat between friends. There is no way as to yet see a list of private chat rooms.

If you are interested in learning more about chats, check with your Internet service provider to see if IRC is available, or whether you will need to Telnet to an IRC client server. If you are a member of a commercial on-line service, check for information on its public and private chat rooms. To experience chat using the World Wide Web, explore these sites.

Time Warner's Pathfinder **http://www.pathfinder.com**
HotWired **http://www.hotwired.com**
The Palace **http:www.thepalace.com**
Globe **http://globe1.csuglab.cornell.edu/global/homepage.html**

Internet Phones

The Internet has made possible the global transmission of text, graphics, sound, and video. Now, a new service has come upon the Internet shore making the real-time transmission of voice possible. New products known as Internet phones let you use your computer as a telephone. Internet phones are the hottest new Internet service to talk with another person anywhere in the world at no more than the cost of your local Internet access. Internet telephones can operate over cable, satellite, and other networks.

However, Internet phones are still in their infancy and not yet a substitute for conventional phones. At this stage in their development, they are still a novelty and far from practical to use as a business tool or for routine communication. To reach another person via the Internet phone, both parties need to be running the same software and be on-line at the same time when the call is made, otherwise the phone won't ring. Currently, most Internet phone software is similar to Internet Relay Chat programs that help users running the same program find and communicate with each other.

Part of the appeal of the Internet phone is the capability to talk to anyone in the world without the cost of a long distance phone call. For the monthly cost of an Internet account two people anywhere in the world can talk for as long and as often as they choose. When one compares this to the cost of national and international phone calls, many are willing to overlook the current limitations and difficulties imposed by this new technology on its users.

The capability and possibilities of the Internet phone have threatened traditional telecommunications companies. The American Carriers Telecommunication Association (ACTA) wants the Federal Communications Commission (FCC) to regulate Internet telephone products. Currently, there are no restrictions on the Net phone, but this could change as hearings are being conducted over the coming months. To keep up to date on these events visit these two Web sites
http://www.von.org or **http://www.netguide.com/net**

How Do I Talk To Someone Using an Internet Phone?

There are two ways that you can communicate with Internet users using Net phones:

- through a central server, similar to an Internet Relay Chat server
- connect to a specific individual by using their IP (Internet Protocol) address

Some Internet users have their own IP addresses; others are assigned an IP address every time they log on. Check with your Internet provider for information on your IP address.

What Do I Need To Use an Internet Phone?

Hardware

Before you can chat using Internet phones you will need the following hardware.

- a sound card for your Macintosh or Windows system
- speakers on your computer
- a microphone for your computer

Sound Card

To have a conversation where both parties can speak at the same time, you will need to have a sound card that supports full-duplexing. Many Macintosh (including the Power Macs) support full-duplex sound. If you are using a PC check your existing sound card. Full-duplex drivers are available if your sound card does not support full-duplexing.

Speakers

The speakers that come with your computer are adequate for the current Net phones. The audio quality of this new technology is not yet what you are accustomed to with traditional telephones.

Microphone

Many computers come with microphones that will be suitable for use with the Internet phones. If you need to purchase a microphone, do not spend more than $10 to $15 dollars on a desktop microphone.

Software

There are several Net phone products that were tested and recommended in the spring of 1996 by *Internet World* magazine.

- VocalTec's Internet Phone
- Quarterdeck Corp's Web Talk
- CoolTalk

Internet Phone (IPhone) was the first Net phone introduced to Internet users in early 1995. After the release and testing of many versions in 1995, the IPhone is considered one of the better Net phones with highly rated sound quality. IPhone is easy to use and resembles the chat environments. When you begin the program you log onto a variety of Internet servers and have the option of joining a discussion group. Once you have joined a group you can call an on-line user by double-clicking on their name. This capability is considered to be one of IPhone's strongest points.

The disadvantage of IPhone is that you cannot connect to a specific individual using their IP address. All connections must be made by first connecting to IPhone's IRC-style servers. Both individuals must be on-line at the same time and connected to the server.

Internet Phone has a free demo version with a one minute talk limit. For more information visit their Web site at **http://www.vocaltec.com** or call (201) 768-9400.

Web Talk is the software program of choice for Internet users with their own IP address. To connect to a specific individual, just enter their IP address. The person you are trying to connect with must also be on-line at the same time. To talk with other on-line users, connect to WebTalk's server. To learn more about WebTalk visit their Web site at **http://www.webtalk,qdeck.com** or call (301) 309-3700.

CoolTalk is distributed by Netscape Communications Corp. and has a cool feature, the whiteboard, that sets it apart from other Internet phone programs. The whiteboard option becomes available after you have connected to another individual. (Connections are made by either logging onto their global server or entering in an individual's IP address.) The whiteboard begins as a blank window. Using standard paint program tools you can enter text, sketch out ideas, draw, or insert graphics. The whiteboard makes this Net phone a most attractive program for Internet business users.

Netscape plans to incorporate CoolTalk with later versions of its Navigator browser. Download a version by connecting to **http://www.netscape.com** or call (717) 730-9501.

NOTE

You can also chat across the Internet using videoconferencing programs such as CU-SeeMe. This program makes it possible for interaction with one individual, small groups, or hundreds in a broadcast. Not only do you hear individuals, but you also can see them in full color on your computer monitor. This program has a whiteboard feature for document collaboration.

CU-SeeMe runs on Windows or Macintosh over a 28.8 modem. If you have a 14.4 modem only, audio is possible. To learn more, visit their Web site at **http://www.cu-seeme.com/iw.htm** or call (800) 241-PINE.

CHAPTER 5
Finding Information and Resources on the Internet

• •

In this chapter, you will learn how to find information and resources on the Internet. You will be using search directories and search engines to find information of interest to you, your career, and your field of study. You will also learn about the following search tools:

- ☛ Yahoo (search directory)
- ☛ Magellan (search directory)
- ☛ Excite (search engine and search directory)
- ☛ Alta Vista (search engine)
- ☛ Infoseek (search engine and search directory)
- ☛ Open Text (search engine)

• •

The Internet contains many tools that speed the search for information and resources. Research tools called "search directories" and "search engines" are extremely helpful.

Search Directories
Search directories are essentially descriptive registries of Web sites. They also have searching options. When you connect to their page, you will find a query box for entering in keywords. The search engine at these sites searches only for keyword matches in the directories' database.

Search Engines
Search engines are different from search directories in that they search World Wide Web sites, Usenet newsgroups, and other Internet resources to find matches to your descriptor keywords. Many search engines also rank the results according to a degree of relevancy. Most search engines provide options for advanced searching to refine your search.

Basic Guidelines for Using a Search Engine

Search engines are marvelous tools to help you find information on the Internet. However, none of these engines delivers consistently accurate and relevant information to your search query, and they provide a high proportion of irrelevant information. Therefore, it is essential that you use several search tools for your research.

Although there are many kinds of search tools, the basic approach to finding information with each is similar:

1. Determine one or more descriptive words (keywords) for the subject you are researching. Enter your keywords into the search dialog box.

2. Determine how specific you want your search to be. Do you want it to be broad or narrow? Use available options to refine or limit your search. Some search engines permit the use of boolean operators (phrases or words such as "and," " or," and "not" that restrict a search). Others provide HELP for refining searches, and some have pull-down menus or selections to be checked for options.

3. Submit your query.

4. Review your list of hits (a search return based on a keyword).

5. Adjust your search based on the information returned. Did you receive too much information and need to narrow your search? Did you receive too little or no information and need to broaden your keywords?

Yahoo

Yahoo is one of the most popular search tools on the Internet and is an excellent place to begin your search. Although Yahoo is more accurately described as a search directory, this Web site has an excellent database with search options available.

Yahoo can be accessed from the Netscape Search Directory button, or by entering this URL **http://www.yahoo.com**

There are two ways to find information using Yahoo: search through the subject index, or use the built-in search engine.

Yahoo Subject Index

When you connect to the Yahoo you will see a list of subjects or directories. Select the topic area that best fits your search needs. Follow the links until you find the information you are searching for.

Using Yahoo To Search For Information

Follow these steps to use Yahoo to search for information:

1. Begin by browsing the subject directory. For example, if you were searching for information on "the use of lasers in medicine," you would first select the *Health* directory and then follow the links to *Medicine*. Explore, and see what is available.

2. Yahoo's search engine can also be used to find information. Enter a descriptive keyword for your subject, one that uniquely identifies or describes what you are looking for. It is often helpful to do a broad search first, though results often provide information on the need to change descriptive keywords or to refine your query.

   Enter the word "laser" (see Fig. 5.1).

3. Click on the **Search** button and review your query results (see Fig. 5.2).

FIGURE 5.1

Yahoo search form and subject index in which the
keyword "laser" has been entered

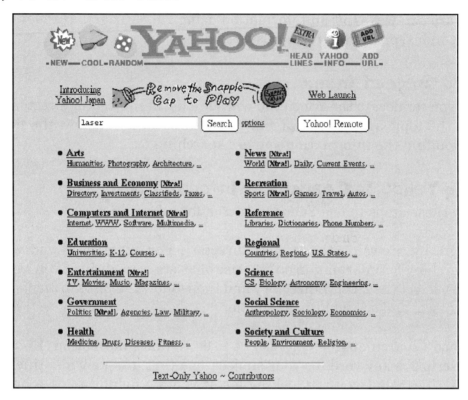

FIGURE 5.2

Yahoo search results from the keyword "laser"

4. You may now want to refine your search. Most search engines have options for advanced searching using boolean logic or more carefully constructed database queries. Review the search page for **Options** or **Advanced Options**. When using Yahoo, click on the **Options** button.

FIGURE 5.3
Yahoo **Options** for refining a search

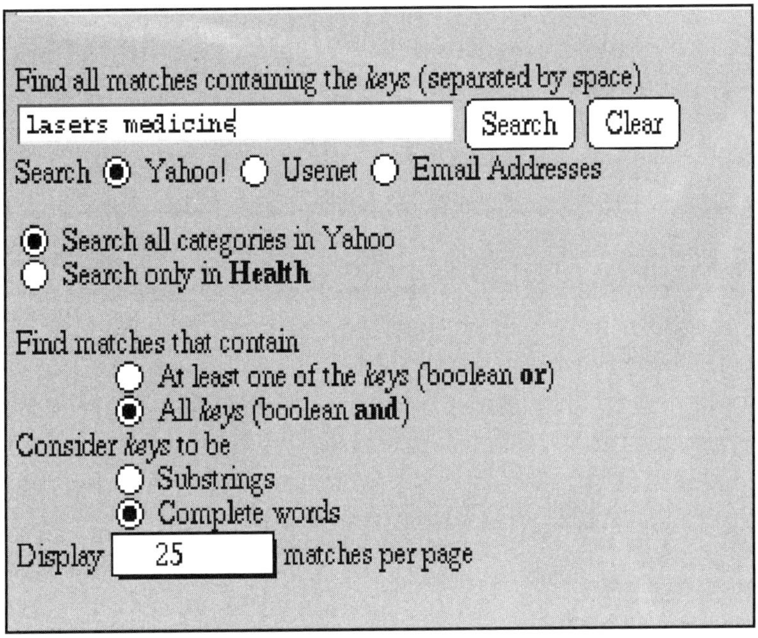

Find all matches containing the *keys* (separated by space)

`lasers medicine`   [Search] [Clear]

Search ● Yahoo! ○ Usenet ○ Email Addresses

● Search all categories in Yahoo
○ Search only in **Health**

Find matches that contain
    ○ At least one of the *keys* (boolean **or**)
    ● All *keys* (boolean **and**)
Consider *keys* to be
    ○ Substrings
    ● Complete words
Display [25] matches per page

If you are using two keywords, do you want Yahoo to look for either word (boolean **or**), both keywords (boolean **and**), or all words as a single string? For example, in the search for "use of lasers in medicine" select boolean **and** because you want to find resources that contain both words "laser" **and** "medicine" in their titles (see Fig. 5.3). Otherwise the search would be too broad and would find all resources that contained either of the keywords "laser" **or** "medicine."

5. Further limit or expand your search by selecting Substrings or Complete words. For example, with your *lasers medicine* search, you would select the search option for *Complete words,* or Yahoo treats the word as a series of letters rather than a whole word. A research

return using substrings would include all incidences where both the words *lasers* and *medicine* appeared in any form.

6. Determine the number of matches you want returned for your search.

7. Submit your query.

8. Review your return list of hits and adjust your search again if necessary.

Magellan

Magellan is another excellent search directory. It provides options for narrowing or expanding your search by selecting sites rated from one to four stars (four stars being the most restricted). You can also restrict your search by excluding sites with mature content by searching for "Green Light" sites only (a green light will be displayed next to the review). **http://magellan.mckinley.com**

FIGURE 5.4
Home page for Magellan with options for specializing your search

Excite

Excite provides the fullest range of services of all the search tools. Excite searches scanned Web pages and Usenet newsgroups for keyword matches and creates summaries of each match. Excite also provides a Web directory organized by category. Excite consists of three services:

- **NetSearch**—comprehensive and detailed searches

- **NetReviews**—organized browsing of the Internet, with site evaluations and recommendations

- the **Excite** Bulletin—an on-line newspaper with reviews of Internet resources, a newswire service from Reuters, and its own Net-related columns

Excite provides two different types of search options: concept-based searching and keyword searching. The search engines described thus far have used keyword search options. Keyword searches are somewhat limited due to the necessity of boolean qualifiers to limit searches.

Concept-based searching goes one step beyond keyword searches—finding what you mean and not what you say. Using the phrase use of "lasers in medicine" a concept-based search will find documents that most closely match this phrase. Excite is available at **http://www.excite.com**

Searching With Excite

1. Type in a phrase that fits your information need.
   Be as specific as you can, using words that uniquely relate to the information you are looking for, not simply general descriptive words. For example, for your laser search enter the following phrase:
   use of lasers in medicine

FIGURE 5.5
Excite Web page displaying concept-based search
using the phrase "use of lasers in medicine"

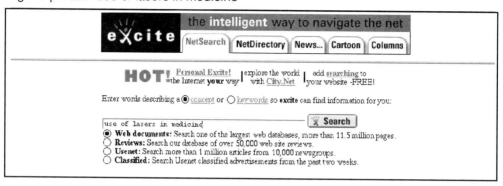

2. If certain words in your search phrase are critically important to the search, you can give them special emphasis by repeating them. For example, if I wanted to find information on the research in the use of lasers in medicine, I would enter in these words:

    research research lasers medicine

    By adding these extra words, the search engine focuses on the double keywords; in this instance the word "research."

3. If you are not sure how to spell a word, type in multiple spellings in your search phrase.

4. There are two ways to have your search results displayed:

    -   **Grouped by confidence**—listed in the order from highest calculated relevance down (Figure 5.6).

    -   **Grouped by site**—shows you where your items in the result list come from (from what physical location).

FIGURE 5.6

Search results for *"use of lasers in medicine"*
(concept-based search) with a percentage
of confidence rating for finding relevant information

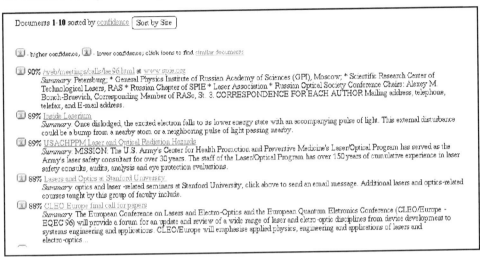

FIGURE 5.7

Excite search using critically important words

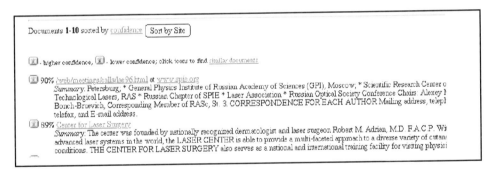

Alta Vista

Digital's Alta Vista is considered one of the best search engines currently available, with one of the largest Web-search databases. Alta Vista's searches are consistently more comprehensive than any of the other search tools. Although you will spend a great deal of time browsing your search

results, you will be provided with as much information as possible on a
search query. **http://altavista.digital.com**

We will use Alta Vista to search for information on "the use of solar energy
for electricity." We will conduct two searches using Alta Vista: a simple
query and an advanced query.

1. A simply query is conducted by entering in keywords or phrases. Do
   not use AND or OR to combine words when doing a simple query. For
   this query we will type in the phrase: *solar energy to produce electricity*.

FIGURE 5.8
Alta Vista Home Page

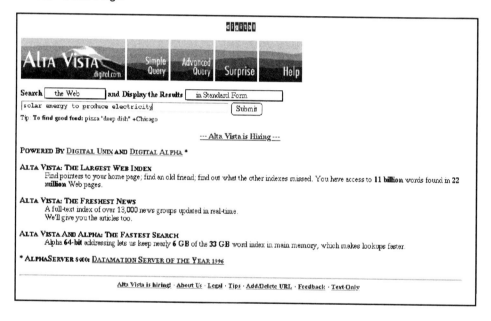

2. Alta Vista's advanced options use the binary operators AND, OR, and
   NEAR and the unary operator NOT. For more information on the
   advanced options, click on the *Help for Advanced Options*.

   We will conduct a more refined search using Alta Vista's advanced
   options. We enter the following words: *solar and energy and electricity
   and produce*.

FIGURE 5.9

Alta Vista advanced options search using the binary operator AND

ALTA VISTA
.digital.com
| Simple Query | Advanced Query | Surprise | Help |

Search [the Web] **and Display the Results** [in Standard Form]

<u>Selection Criteria:</u> Use only Advanced Query Syntax with **AND, OR, NOT** and **NEAR**. Simple Query Syntax will not work!

```
solar and energy and electricity and produce
```

<u>Results Ranking Criteria:</u> documents containing these words will be listed first. If left blank, the matching documents will not be sorted.

Start date: [   ]   End date: [   ]   e.g. 21/Mar/96

[Submit Advanced Query]

Infoseek

InfoSeek is a professional service provided by InfoSeek Corporation. In 1995, InfoSeek introduced their easy-to-use search services to subscribers for a monthly subscription fee. Because of its popularity, the services have been expanded to include two new options: Infoseek Guide and Infoseek Professional.

Infoseek Guide is the free service that integrates the latest search technology with a browsable directory of Internet resources located on World Wide Web sites, Usenet newsgroups, and other popular Internet resource sites. Users can choose to use the search engine and enter keywords or phrases, or browse the navigational directories.

Visit the InfoSeek Guide site and try these tools for finding Internet information and resources. **http://guide.infoseek.com**

Infoseek Professional is a subscription-based service that offers individuals and business professionals comprehensive access to many Internet resources such as newswires, publications, broadcast programs, business, medical, financial and government databases. The difference between Infoseek Guide and Professional is the capability to conduct more comprehensive searches and to have options for refining and limiting your searches. For example, you can conduct a search query by just entering in a question such as "How can I get information on ISDN?" You can also limit your query to just the important words or phrases that are likely to appear in the documents you are looking for:

information on the best "ISDN" "hardware"

By identifying the key words or phrases (**ISDN** and **hardware**) with quotes, your search accuracy is greatly enhanced.

Professional offers a free trial period. To learn how to perform the most efficient search, link to Infoseek's information on search queries and examples. **http://professional.infoseek.com**

FIGURE 5.10

Infoseek guide for information and resources

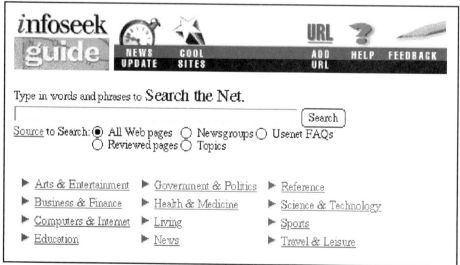

Open Text

Open Text has one of the most comprehensive collections of search tools and is one of the best designed search engines on the Internet.
http://www.opentext.com

Open Text offers many search options:

- simple query on words;
- a power search using up to five operators between terms (*and*, *or*, *not*, *but not*, *near*, and *followed by*);
- options to create your own weighted search;
- results scored by relevancy; and
- an option to show a report of where Open Text found your search matches.

Open Text produces better returns on your search if you break up a phrase into keywords. For example, when "use of solar energy to produce electricity" was entered, Open Text reported "no matches." When the query was changed into individual search terms "solar energy" and "electricity" a large number of results were displayed.

FIGURE 5.11
Open Text page showing query entered with
two search terms: "solar energy" and "electricity"

Search Tips...

- When you find a resource that you may want to return to, make a bookmark or add the page to your hotlist.

- Avoid using common words such as articles and propositions in your searches. These words are frequently ignored by search engines. Use words that describe or are very specific to your search topic.

- Use boolean qualifiers such as *"and"* or *"or"* to limit your search.

- If your search result is too limited using boolean qualifiers, re-submit the search with these qualifiers.

- Evaluate your use of keywords and the types of results you obtain using different search engines.

- Look for keywords displayed with your search results that better describe your topic. Re-submit your search using these new keywords.

- Each search engine provides a slightly different type of search. For one type of search, one engine might be more appropriate than another. For example, in the past you may have used different resources for your searches such as an encyclopedia, periodicals, or books. Each resource is good for a different type of information. The same is true with search engines.

- Use more than one search engine because each search site has its own database resources. Some sites may have information that others do not.

- Make a bookmark for each of the search engines described here. Add others that you might find useful.

PART II
The Web and Electronic Technology

CHAPTER 6
Cool Electronic and
Electrical Web Sites

. . . .

In this chapter, you will find a plethora of electronic resources on the Internet. Explore how Electronics Technology is using the Internet. Categories include:

- ❥ Links to Electronic and Electrical Sites
- ❥ Technical Societies and Publications
- ❥ Circuit Archives, Technical Information, and Projects
- ❥ Universities and Government Sites
- ❥ Electronics Companies
- ❥ Electronic Listserv Mailing Lists
- ❥ Electronic Usenet Newsgroups

. . . .

Links to Electronic and Electrical Sites

Journey into the cyberworld of Electronic Engineering and explore many of the best electronic and electrical sites.

Best Electrical Engineering Site: EE/CS Mother Site
The EE/CS Mother Site is a starting point when looking for Electrical Engineering and Computer Science related information. It is sponsored by the Stanford IEEE. Companies and product information are listed, **http://www-ee.stanford.edu/soe/ieee/eesites.html**

Cool Electronics Stuff
Links to electronic resources.
http://pasture.ecn.purdue.edu/~laird/Electronics/index.html

Circuit World Online Services

This provider of information and commercial services to the Electronics Manufacturing Industry has more than 500 links. Major categories include Electronic Assembly and Packaging, Printed Circuit Fabrication, Electronic Components and Semi-Conductors, Design and Engineering, Industry Associations, OEM Manufacturers and much more. This comprehensive site takes you through the entire spectrum of electronics. **http://www.circuitworld.com**

Electronic & Computing Links

This site, maintained by Ben Shirley, School of Electronics, UCS, has links to sites that he has found useful or fun. **http://www.ucsalf.ac.uk/~bens/eleclink.htm**

Yahoo

Learn about how your electronics education can open doors to the future. At Yahoo's Engineering Web page, you will find links to aerospace, agricultural, automotive, biomedical, chemical, electrical, environmental, industrial, mechanical, naval, nuclear, optical, petroleum, reliability, software, and structural and welding engineering. **http://www.yahoo.com/Science/Engineering**

Technical Societies and Publications

These sites offer the new engineer an opportunity to join with other engineers to share knowledge and discuss real-world problems. Trade magazines and newsletters keep abreast of constant changes in the world of electronics and electrical engineering.

CAD Electronics

A newsgroup with lots of articles relating to CAD in electronics. Links to shareware CAD, RF Mosfet Models, and Document Conversion Resource Center are some of the subjects covered. **news:sci.electronics.cad**

Computing Research Association

An association of more than 150 North American academic departments of computer science and computer engineering as well as industrial laboratories. **http://cra.org**

The Directory of Engineering and Scientific Trade Technical Magazines

One of the most useful sites to locate trade publications on any technical subject. Links are provided to those who have online sites and addresses.
http://www.techexpo.com/tech_mag.html

A Directory of Engineering and Science Societies and Organizations

Links to and addresses of technical and scientific organizations of every imaginable subject. A great place to explore.
http://www.techexpo.com/tech_soc.html

EDN Magazine

This technical magazine is published every two weeks and keeps electronics professionals up-to-date on the latest products, design techniques and emerging technologies. View the latest issue on line or subscribe on-line to receive your own copy. **http://www.ednmag.com**

EE Times

The latest news for electrical and electronic engineers.
http://techweb.cmp.com/techweb/eet/current

EIA Electronics Industries Association

An established trade organization representing U.S. electronics manufacturers. It is a multifaceted group with interests in technical standards, market analysis, government regulations, trade shows, and seminar programs. **http://www.eia.org**

Electronics & Electrical Engineering Laboratory

A government lab to promote standardization in many fields of electronics and electrical engineering. **http://www.nist.gov/item/About_NIST_Electronics_and_Electrical_Engineering_Laroratory.html**

Institute of Electrical and Electronics Engineers (IEEE)

Institute of Electrical and Electronics Engineers, Inc. The world's largest technical professional society. Its mission is to promote the development and application of electrotechnology. It provides Technical Societies in 37 fields of special interest ranging from Aerospace to Vehicular Technology. Technical conferences are listed as well as the IEEE Bookstore for all your technical publication needs. Industry Standards for the electronic industry also can be ordered. The Student Activities Committee's Home Page will be of interest to students and educators alike. Membership information and student web pages are listed. The latest in fashion wear, the IEEE T- shirt, can be the smartest addition to your wardrobe.

IEEE Publications includes the on-line *Spectrum* magazine and *The Institute*, the IEEE monthly newspaper. Information to become a Professional Engineer, help with career planning, self study courses, teaching opportunities, women in engineering home page, and much, much more make this a MUST VISIT site. **http://www.ieee.org**

ISO International Organization for Standardization
The premiere group promoting standardization in the world today consisting of a federation of national and local organizations from more than 100 countries. **http://www.iso.ch/welcome.html**

ITU International Telecommunication Union
An organization headquartered in Geneva, Switzerland that coordinates global telcom networks and services for the government and the private sector. **http://www.itu.ch**

PC/104 Magazine
This on-line journal of Controlled Systems focuses on PC/104 embedded control systems and the products that make them work. Also check out their jobs database for something of interest. You can also fill out a form to list yourself as wanting a job. **http://www.controller.com/pc104**

Professional Organizations and Government Labs of Interest to Electrical Engineers
A site maintained by the Department of Electrical Engineering at the University of Missouri–Rolla. **http://www.ee.umr.edu/orgs**

Stanford Center for Integrated Systems
A cooperative venture between Stanford University and member industrial firms. **http://snf.stanford.edu/cis**

Telecommunications—The Telecommunications Library
Here is a wealth of information about the telecommunication field
sponsored by LDDS WorldCom Network. Visit the Telecom Digest, one of
the oldest and most respected forums on the Internet. In its 12 year history,
the Digest has covered every imaginable issue of telecommunications.
Telecomreg is the source for the latest on regulation in the
telecommunications industry. Academic papers are published thru the
Research Institute for Telecommunications and Information Marketing
(RITIM). The Insight Research Corporation provides trend analysis and
comparative market research in the wireless, voice, data, and video
communications field. In addition, jobs that are available at LDDS are
published. An excellent site for engineers interested in the
telecommunications field. **http://www.wiltel.com/library.html**

Circuit Archives, Technical Information and Projects

Every engineer needs reference material and never has enough. These
sites offer thousands of circuits and technical information to construct
your own projects and research many subjects. This is the great advantage
of the Web because new technology is being added daily and it is there
for the taking. Log on now!

Beast 95
Bucknell Engineering Animatronics Systems Technology is an ongoing
senior design project with the goal of creating a fully interactive
animatronic figure. This figure is interactive, which means it can sense,
for instance, how many people are in the room and modify its performance
based on that data, such as begging you to stay. A block diagram of the
animatronic system shows an overview of the microcontrollers and
sensors. The components used to build the Beast are listed along with
photos and details of the project. Beast 96 has already been started so
check it out! **http://www.eg.bucknel.edu/~beast96**

Berkeley Sensor & Actuator Center
Learn about the latest in sensor and actuator technology.
http://www-bsac.eecs.berkeley.edu

The Center for Compound Semiconductor Microelectronics

Funded by the National Science Foundation, this University of Illinois facility seeks to address research in optoelectronic integrated circuits. New high-speed communications and data processing technology is a goal of the group. **http://www.ccsm.uiuc.edu/ccsm**

Chaos

Since the mid-1970s, the Chaos Group at Maryland has done extensive research in various areas of chaotic dynamics ranging from the theory of dimensions, fractal basin boundaries, chaotic scattering, and controlling chaos. It is hoped that their knowledge will be useful to others. **http://www-chaos.umd.edu/chaos.html**

Circuit Archive

This circuit archive at the University of Washington site which contains many circuits for applications including IR related circuits, PC circuits, telephone circuits, and miscellaneous circuits. You are encouraged to submit your own designs to the archive. **http://weber.u.washington.edu/d99/pfloyd/ee**

The CPU Info Center

This site contains various CPU related information and with special focus on CPU architecture. Comparisons between types of CPUs are given. Technical papers, embedded microprocessor information, the history of CPUs and a die photo gallery can be found here. **http://infopad.eecs.berkeley.edu/CIC/about.html**

Electrical Engineering Shop

Lots of unusual electronics projects that have not been seen on the Web. Everything you could want to know about PC Serial Communication, signals used on ISA Buss, a Binary-to-IntelHex Conversion utility, and prototyping tips and pitfalls are listed at this University of Nebraska server. **http://engr-www.unl.edu/ee/eeshop/miscinfo.html**

The Electronic Cookbook Archive

Sponsored by the University of Alberta's Department of Electrical Engineering , this circuit "cookbook" archive covers audio, computers,

digital, power, RF, software, telecommunications, video and wave shaping.
http://www.ee.ualberta.ca/html/cookbook.html

Electronics In Music

A music electronics archive containing circuits for many electronic music effects. Delay based effects, distortion, overdrive, and fuzz-tones are some of the circuits shown. A very interesting site for the musically inclined electronics engineer.
http://rowlf.cc.wwu.edu:8080/~n9343176/schems.html

How Semiconductors Are Made

A step by step look at the manufacturing process used to make semiconductors.
http://rel.semi.harris.com/docs/lexicon/manufacture.html

Lexicon of Semiconductor Terms

A site where you can look up an unfamiliar technical semiconductor term and find out its meaning.
http://rel.semi.harris.com/docs/lexicon/preface.html

MICAS: Medical and Integrated Circuits and Sensors
http://www.esat.kuleuven.ac.be/micas

Microelectronics

Engineers interested in microelectronic circuit design will want to check out the newsletters at this University of Tennessee site. Subscribers are notified when a new issue is published. Integrated circuit prototyping via MOSIS, as well as microelectronic systems, are discussed.
http://microsys6.engr.utk.edu:80/ece/msn

Microelectronics Research Center (MRC)

A NASA Space Engineering Research Center for VLSI, ASIC, and device modeling. **http://www.mrc.uidaho.edu:80/**

NASA Electronic Packaging and Processes Branch

This site at the Goddard Space Flight Center is set up to implement new packaging technologies and address problems in older technologies.
http://package.gsfc.nasa.gov/package.html

The National Nanofabrication Users Network
With the motto "No Job Too Small," this site is devoted to share research in nanoscale science. **http://snf.stanford.edu/NNUN**

Ohm's Law
Check out the Ohm's Law experiment from California State. You can make measurements directly on equipment linked to their server. **http://plabpc.csustan.edu/physics/expt/ohmslaw.htm**

Optoelectronic Computing Systems Center
Their mission of the Colorado Advanced Technology Institute is to provide cross-disciplinary research and education in optoelectronic technology. **http://ocswebhost.colorado.edu**

QuestNet
A site where the engineer can search for information on semiconductor and integrated circuit devices. **http://www.questlink.com**

Research In Analog IC Design
University of California–Berkeley projects and research with a focus on the design of analog circuits for high integration. A monolithic CMOS RF Transceiver is described as well as High Speed, Low Power CMOS ADCs. **http://kabuki.eecs.berkeley.edu**

Robotics
Links to robotics. **http://www.yahoo.com/Science/Engineering/ Mechanical_Engineering/Robotics**

Semiconductor Subway
The Semiconductor Subway site maintained by the Massachusetts Institute of Technology. Links to other sites with a focus on semiconductor related sites. Upcoming conferences relating to semiconductor manufacturing can be found. **http://www-mtl.mit.edu/semisubway.html**

Signal Processing Information Base (SPIB)
Digital information including data papers, software, newsgroups, and

bibliographies that are relevant to signal processing and research.
http://spib.rice.ecu:80/spib.html

Signal Processing URL Library
Links to web sites with a focus on signal processing.
http://www-dsp.rice.edu/splib

Solid State Lab
The University of Michigan department of Electrical Engineering and Computer Science Solid State Electronics Laboratory. The latest technology for semiconductor design, including materials, devices and integrated circuits is discussed. Ongoing projects are reviewed, as well as group efforts between students, technicians, and engineers.
http://www.eeecs.umich.edu/dp-group

Spreadsheets
A number of spreadsheets dealing with various aspects of teaching basic electricity have been compiled. Click "batteries" for a spreadsheet containing graphics that tea about batteries in series. "Resistors" provides practice for computing resistors in series. Click" circuits" for a spreadsheet that uses simple circuits to teach about Ohm's Law.
http://192.239.146.18/SS/Spreadsheets.html

The Tech
A hands on museum of technology located in the heart of Silicon Valley. Explore this site on your next trip to San Jose or logon the site and explore from your computer. **http://www.thetech.org/about.html**

Theremin Home Page
What is a Theremin? A theremin is an electronic musical instrument based on the the theory of beat frequencies. Learn more about theremins by visiting this interesting Web site.
http://www.ccsi.com/~bobs/theremin.html

Turbulence Links Around The Web
Here are some links to a variety of turbulence modeling and related WWW sites. **http://stimpy.ame.nd.edu/gross/fluids/turbulence.html**

VLSI Engineering

VLSI design, free CAD shareware, conferences—and best of all—job opportunities are here. Interesting links to other VLSI educational projects and educational courses in VLSI are listed at this University of Idaho site. **http://www.mrc.uidaho.edu/vlsi/vlsi.html**

Weird Science

"It is not uncommon for engineers to accept the reality of phenomena that are not yet understood, as it is very common for physicists to disbelieve the reality of phenomena that seem to contradict contemporary beliefs of physics." - H. Bauer **http://www.eskimo.com/~billb/weird.html**

Universities and Government Sites

Many universities and colleges have sites maintained by the Department of Electrical and Electronic Engineering. Teaching opportunities, graduate and post graduate programs, research projects, and courses are listed. Government sites also have research programs and provide the direction for implementation of Standards in the Electronic and Electrical Industry.

Alabama Microelectronics Science and Technology Center (AMSTC)
http://www.eng.auburn.edu/department/ee/amstc/amstc.html

CalTech Department of Electrical Engineering
http://electra.micro.caltech.edu

Carnegie Mellon Department of Electrical Engineering
http://www.ece.cmu.edu

Cornell University School of Electrical Engineering
http://www.ee.cornell.edu

Duke University Department of Electrical and Computer Engineering
http:/www.ee.duke.edu

Electrical Engineering Programs

Listed here are links to home pages of Electrical Engineering academic programs throughout the world. A great way to see what's happening in academic electronics world wide.
http://www.ee.umr.edu/schools/ee_programs.html

Johns Hopkins University Electrical and Computer Engineering Department
http://www.ece.jhu.edu

Kansas State Department of Electrical and Computer Engineering
http://www.eece.ksu.edu

Companies

The bulk of the job opportunities are with private industry. Companies that would be of interest to electrical and electronic engineers are listed. Many have employment pages where job openings are listed. Some have the capability for you to submit your résumé on-line. This is a great way to learn about a specific company or industry before your job interview.

Allied Signal

An 88,000 employee company that services the aerospace, automotive, chemical, and advanced materials industries. Many opportunities are available to the electrical/electronic engineer in a company as diverse as this one. Check out the site for an overview of its activities.
http://www.alliedsignal.com
e-mail: address not given

AT&T

This company is one of the oldest and best known in the telecommunications field. AT&T's Bell Telephone Labs is a premiere company devoted to research and innovation and has distinguished itself with seven Nobel prize scientists. While visiting this site review the job links for college students. **http://www.att.com**
e-mail **webmaster@att.com**

Compaq Computer Corp.

This well-known computer manufacturer has supplied products to the computer industry for many years. Visit the site to see what's new at Compaq. **http://www.compaq.com**
e-mail: **webmaster@compaq.com**

Dallas Semiconductor

A diversified manufacturer of silicon products such as digital thermometers, microcontrollers, silicon-timed circuits, and more. Check out employment opportunities at the site and submit your résumé.
http://www.dalsemi.com
e-mail: **recruiter@dalsemi.com**

Delco Electronics

A worldwide manufacturer of automotive and consumer-related products. With 28,000 employees, many opportunities are available for the electronic and electrical engineer.
http://www.delco.com/recruitment.html#employment
or view the home page at:
http://www.delco.com
e-mail: address not given

Digital Equipment Corporation

Worldwide supplier of network computer systems, software, and services. This site provides career opportunities for students in the electronic engineering field. **http://www.dec.com**
e-mail: **jobs-us-servers@digital.com**

Dolby Laboratories

This well-known maker of audio products has some interesting career opportunities for the electronic engineer. Visit the career page:
http://www.dolby.com/carops.html
and check out the company page at:
http://www.dolby.com
e-mail: address not given

Fujitsu

This high technology electronics manufacturer of drives, mobile telephones, radios, and other equipment offers many job opportunities in the United States and abroad. Check out the job site of your choice.
http://www.fujitsu.com
e-mail: **webmaster@fujitsu.com**

General Electric

General Electric Company is a worldwide diversified technology, manufacturing, and services company. **http://www.ge.com**
e-mail: **good.things@corporate.ge.com**

General Motors

Excellent opportunities are here at the world's largest auto manufacturer. Career opportunities in their engineering matrix chart list divisions where electrical and electronic engineers would fit. While your there check out this huge site and its links to various GM divisions as well as taking a Virtual Reality trip through the newest cars.
http://www.gm.com/edu_rel/careers.htm#Engineering
visit the corporate site at: **http://www.gm.com/index.htm**
e-mail: address not given

GTE

One of the largest telecommunications companies in the world with $20 billion in sales. Many opportunities exist for the recent college graduate as well as the seasoned professional. Check out the recruitment site at;
http://www.gte.com/Working/Campus/campus.html
or look at the corporate site at:
http://www.gte.com
several e-mail addresses for human resources sites can be found at:
http://www.gte.com/career/contact.html

Harris

A worldwide company doing business in electronic systems, semiconductors, communications, and Lanier Worldwide Office Systems. With 27,000+ employees, many job opportunities exist for the college graduate. **http://www.harris.com**
e-mail: **webmaster@harris.com**

IBM
An international computer manufacturer. IBM has an excellent employment page at: **http://www.empl.ibm.com** and a lot of information about the company and its programs: **http://www.ibm.com**
e-mail: **askibm@info.ibm.com**

Intel
Logon to Intel for a look at the company's latest technologies. Also drop off your résumé at its Employment Office. **http://www.intel.com/intel**

Lockheed Martin
The combined companies of Lockheed and Martin Marietta, best known for their aerospace history, offer many opportunities for the aspiring engineer. **http://www.lmco.com**
e-mail: **webmaster@lmco.com**

Motorola
A $27 billion company with job opportunities worldwide. Visit the job matrix to see where you might like to work and follow up with a link to the business unit of your choice. A leading provider of semiconductors, cellular telephones, two-way radios, automotive products, and much more.
http://www.motorola.com/UR/intro1.html
For overall information visit the corporate site at:
http://www.mot.com
e-mail: **webmaster@mot.com**

Motorola Semiconductor Products Group
Find the latest product information from the Analog Microcontroller Division, Analog IC Division, MOS Digital–Analog IC Division, RF Products–Communication Semiconductor Division, Digital Signal Processors (DSP).
http://design-net.com/home/prodgroups/html/prod_groups.html

Okidata Corp.
A manufacturer of semiconductors with job opportunities for the college graduate. **http://www.okisemi.com/index.html**
e-mail: **webmaster@obd.com**

Panasonic - Matsushita Electric

This $7 billion company employs over 16,000 people in 21 manufacturing sites in North America alone. The range of products manufactured is exceptionally wide with a focus in consumer electronics.

http://www.mitl.research..panasonic.com/pana.html
e-mail: **webmaster@research.panasonic.com**

Philips Semiconductors

The tenth largest supplier of semiconductors in the world in wireless communications, micro controllers, audio, video, and more. Employment information is at:

http://www.semiconductors.philips.com/ps/philips19.html
Visit the corporate site at:
http://www.semiconcuctors.philips.com
e-mail: **webmaster@semiconductors.philips.com**

Texas Instruments

A supplier of electronic and scientific calculators, printers, notebook computers, electrical controls, etc.

http://www.ti.com
e-mail: address not given

Videonics

A video editing company with opportunities listed at this Web site.

http://www.videonics.com/employment.html
e-mail: **helpline@videonics.com**

Listserv Mailing Lists for Electronic Technology

Chemical Engineering

Discussion of Interfacial Phenomena
Mail to: LISTSERV@WSUVM1.CSC.WSU.EDU

Engineering

Discussion of engineering and construction
Mail to: MAILBASE@MAILBASE.AC.UK

Nuclear Engineering

Discussion of nuclear energy, research, and education
Mail to: LISTPROC@MCMASTER.CA

Technology

I-TV

Discussion of two-way interactive television used for education and community development
Mail to: LISTSERV@KNOWLEDGEWORK.COM

INFO-FUTURES

Discussion of the effect of technolgy in industry
Mail to: INFO-FUTURES-REQUEST@WORLD.STD.COM

PHOTO-CD

Kodak CD products and technology
Mail to: LISTSERV@INFO.KODAL.COM

SATNEWS

Satellite television industry newsletter
Mail to: SATNEWS-REQUEST@MRRL.LUT.AC.UK

For information on finding listserv mailing lists see page 45.

Usenet Newsgroups for Electronic Technology

Electronics Newsgroups

sci.electronics

This newsgroup evolved from the *sci.electronics.basic* group in January of 1996.

sci.electronics.basics

A forum for the discussion of electronics, where there is no stupid question. A place to ask elementary questions about electronics

sci.electronics.cad

A forum for the discussions of Computer Aided Design software for use in designing electronic circuits and assemblies.

sci.electronics.components

Discussions of integrated circuits, resistors, capacitors.

sci.electronics.design

Discussions on electronic circuit design.

sci.electronics.equipment

Information on test, lab, and industrial electronic products.

sci.electronics.misc

General discussion of the field of electronics.

sci.electronics.repair

A forum for discussing the fixing of electronic equipment.

misc.industry.electronics.marketplace Electronics products and services.

Other Electronic Technology Newsgroups

alt.cad
alt.cad.autocad
alt.electronics.analog.visi
alt.energy.renewable
alt.sustainable.agriculture
alt.solar.photovoltaic
alt.solar.thermal
clari.tw.aerospace
comp.cad.autocad
comp.cad.microstation
comp.robotics.misc
comp.robotics.research
sci.bio.misc
sci.bio.technology
sci.chem
sci.energy
sci.engr
sci.engr.biomed
sci.engr.chem
sci.engr.civil
sci.engr.heat-vent-ac
sci.engr.lighting
sci.engr.manufacturing
sci.engr.mech
sci.engr.semiconductors
sci.engr.television.advanced
sci.engr.television.broadcast
sci.environment
sci.geo.satellite-nav
sci.life-extension
sci.materials
sci.med.physics
sci.military.naval
sci.nanotech
sci.optics
sci.optics.fiber
sci.research

CHAPTER 7
Using Cyberspace for Career Planning

Today more and more career development centers are using the Internet as a resource for career planning. Major career planning activities include self-assessment and career exploration. In this chapter, you will

- ➡ take a self-awareness journey to learn more about yourself and your personal and professional needs;

- ➡ take a journey into cyberspace to research jobs that fit you as a person; and

- ➡ learn how to use the Internet for career exploration: communication with people, electronic publications, career resources, and professional services.

Self-awareness Journey

Self-assessment is the first step in career planning. Self-assessment is an important process that requires inner reflection. The goal of this reflective process is to help you develop a better understanding of your interests, talents, values, goals, aptitudes, abilities, personal traits, and desired lifestyle. You will use this information to help find a job that fits you as a person. This personal survey is very important in helping you become aware of the interrelationship between your personal needs and your occupational choices.

Start by identifying:

- your interests and what is important to you;
- what you enjoy doing in your free time;
- what skills you learned in the classroom or from an internship that are related to your career interests;

- your accomplishments;
- abilities and capabilities;
- work experience that you have had related to your career interests;
- personal traits and characteristics;
- your strengths and weaknesses; and
- physical and psychological needs.

Ask these questions regarding career considerations:

- Where would you like to live? In a city, suburbs, the country, the seashore, or the mountains?

- Is there a specific geographic location where you would like to live?

- How do you feel about commuting to work? Would you drive a long distance to work for the advantage of living outside of a city?

- Is the community that you live in important? For example, do you value a community that is outdoor oriented or family oriented?

- What type of work environment is important to you? Do you want to wear a power suit every day or be casual?

- Is making a lot of money important to you?

- How do you feel about benefits and promotion options?

- Are flexible hours and free weekends important? For example, do you value free time to exercise and participate in outdoor activities? Are you willing to sacrifice this part of your life for a job? Would you be satisfied with making enough money to live on and have more free time?

- Do you mind working long hours each day or weekends? How do you feel about taking vacation time?

- Do you want to work for a large or small company? Would you rather work for a small company where everyone knows each other and the atmosphere is perhaps a little more casual? Or is it more important to be with a large company with many career advancement opportunities?

- Where would you like to be in your professional life in 5 years? 10 years? Does this company offer advancement opportunities that fit your goals?

- How do you feel about work-related travel? Do you mind traveling if a job requires you to do so? Do you mind giving up portions of your weekend to travel? How many days a month are you willing to be away from home?

- How do you feel about being a member of a work team?

After you have completed your self-awareness journey, you are ready to use this information to explore career options.

Career Exploration

The goal of career exploration is to help you to find job opportunities that match your personal and professional needs. Career exploration involves gathering information about the world of work. You will eliminate or select jobs based on what you learned in your self-assessment. For example, if you determine that location is an important factor when selecting a job, you would use this criteria to select or eliminate job opportunities based on where a particular company is located. Information about the work environment and corporate culture will be more difficult to obtain.

There are many ways to obtain information about the world of work. In this section we will explore several options involving communication and the use of the Internet to acquire information.

People As Information Resources

Internships and work experience provide excellent opportunities to learn about companies and their world of work. For example, if you are doing an internship for a company observe the work ethic and corporate environment. Ask someone doing a job that interests you what it is like to work for the company. How many hours do they work per day? Are they expected to work weekends? How does their department, boss, and other employees view vacations? Do they have free time during a day for personal interests such as running, cycling, or working out at the gym? What are the company's expectations of its employees? If a job position requires the employee to travel, ask how many days per month they travel? When do they leave to travel; when do they return? How are they compensated for overtime?

Other sources for obtaining information from people include

- talking with your career counselors;
- attending seminars and workshops where you can interact with professionals and ask questions;
- attending conventions and job fairs;
- joining a professional organization; and
- NETWORK, NETWORK, NETWORK!

Publications As Information Resources

Professional publications provide valuable information about the world of work. Check with your career counselor or professors for publications that will provide useful information. See the electronic links to technical societies and publications beginning on page 73. Another Internet resource is the electronic newsstand. This excellent Web site has links to many electronic versions of publications.

http://www.enews.com

The Internet As an Information Resource

The Internet has many valuable resources for learning about the world of work. Resources include:

- World Wide Web sites of companies
- Usenet newsgroups
- listserv mailing lists
- job and career resources on the Internet

World Wide Web

Many companies have World Wide Web sites. You will find many of these Web sites useful in learning about a company's products and services and, in some instances, about their work environment. Use search engines described in Chapter 5 to enter in the name of companies you are interested in to help find their Home Page. Also enter in keywords descriptive of the type of job you would like.

Listed below are several URLs that have links to electronics and manufacturing companies.

http://www.ctrl-c.liu.se/other/admittansen/netinfo.html

http://www.scescape.com/WorldLibrary/business/companies/elec.html

Usenet newsgroups

In the virtual community of the Internet, Usenet newsgroups are analogous to a cybercafé where people with similar interests gather from around the world to interact and exchange ideas. Usenet is a very large distributed bulletin board system (BBS) that has several thousand specialized discussion groups. Currently there are over 20,000 newsgroups with about 20 to 30 more added weekly. Anyone can start a newsgroup.

You can subscribe to a newsgroup, scan through the messages, read messages of interest, organize the messages, and send in your comments or questions.

> ## NOTE
> Your college or university must carry Usenet News before you can use your Internet browser to read and interact with newsgroups.

Listed below are several Usenet newsgroups that are relevant to job searching and career planning:

> **misc.jobs.offered**
> **misc.jobs.offered.entry**
> **misc.jobs.contract**
> **misc.jobs.resumes**
> **misc.jobs.misc**

Netscape Navigator supports Usenet newsgroups. To view all the newsgroups available on your college or university network, follow these steps from within Netscape Navigator 2.0:

1. Click on the **Window** pull-down menu.
2. Select **Netscape News**.
3. Within the Netscape News window, go to the **Options** pull-down menu.
4. Select **Show All Newsgroups**.

Visit this Web site for a listing of Usenet newsgroups. **http://ibd.ar.com/ger/**

Visit this Web site and use a simple search tool to locate Usnet newsgroups of interest. **http://www.cen.uiuc.edu/cgi-bin/find-news**

For more information on using Netscape for reading newsgroups, refer to *Netscape Adventures—Step-By-Step Guide To Netscape Navigator and the World Wide Web*.

Listserv Mailing Lists

A *listserv* is the automated mailing system that distributes electronic mail. Mailing lists provide a forum where individuals of shared interests can exchange ideas and information; any members of the group may participate in the resulting discussion. This is no longer a one-to-one communication like your e-mail, but rather a one-to-many communication. Electronic mail written in the form of a report, article, abstract, reaction, or comment is received at a central site and is then distributed to the members of the list.

Finding a Listserv for Jobs and Career Planning

There are several Internet resources to help you to find a listserv mailing list for jobs or career planning. A World Wide Web site for mailing lists is:

http://www.tile.net/tile/listserv/index.html

Travel to this excellent Gopher server and follow the path to information on current mailing lists. You can also do a search for mailing lists by subject.

     Gopher: **liberty.uc.wlu.edu**
     path: Explore Internet Resources/Searching for Listservs

You can also use electronic mail to request information on listserv mailing lists on a particular topic. Send an e-mail message to:

LISTSERV@vm1.nodak.edu

In the message body, type: **LIST GLOBAL / *keyword***

For example, if you were looking for a mailing list on jobs you would type in the message body: **LIST GLOBAL/ jobs**

TIPS for New Users of Newsgroups and Listservs

Tip 1...

After you subscribe to a list or newsgroup, don't send anything to it until you have been reading the messages for at least one week. This will give you an opportunity to observe the tone of the list and the type of messages that people are sending. Newcomers to lists often ask questions that were discussed at length several days or weeks before.

Tip 2...

Remember that everything you send to the list or newsgroup goes to every subscriber on the list. Many of these discussion groups have thousands of members. Before you reply or post a message read and review what you have written. Is your message readable and free from errors and typos? When necessary, AMEND BEFORE YOU SEND.

Tip 3...

Look for a posting by someone who seems knowledgeable about a topic. If you want to ask a question, look for their e-mail address in the signature information at the top of the news article. Send your question to them directly rather than to the entire newsgroup or listserv.

Tip 4...

Proper etiquette for a mailing list is to not clog other people's mail boxes with information not relevant to them. If you want to respond to mail on the list or newsgroup, determine whether you want your response to go only to the individual who posted the mail or you want your response to go to all the list's subscribers. The person's name and e-mail address will be listed in their posting signature.

Tip 5...

The general rule for posting a message to a list or newsgroup is to keep it short and to the point. Most subscribers do not appreciate multiple page postings.

If you are contacting an individual by electronic mail, identify yourself, state why you are contacting them, and indicate where you found their posting. Again, be as succinct and to the point as possible.

Request further information by either e-mail or by phone.

Job And Career Resources On the Internet

There are many excellent career planning and job-related resources on the Internet. Listed below are a few Web sites to investigate.

- **CareerMosaic** — Follow the links to the Resource Center
  http://www.careermosaic.com

- **Career Magazine**
  http://www.careermag.com/careermag

- **Monster Board Resource Link**
  http://199.94.216.76:80/jobseek/center/cclinks.htm

- **Occupational Outlook Handbook**
  http://www.jobweb.org/occhandb.htm

- **Online Career Center**
  http://www.occ.com

▣ **Riley Guide**
http://www.jobtrak.com/jobguide/what-now.html

▣ **US Industrial Outlook** — information on job market realities
http://www.jobweb.org/indoutlk.htm

Professional Services As Information Resources

One valuable service to job seekers who want to learn what it's really like to work at a specific company or within a specific industry is Wet Feet Press. This service provides comprehensive in-depth analyses of companies at a cost of $25 per report. If you are currently enrolled as a Bachelor's or Master's student at a Wet Feet Press "Information Partnership" university, your cost is only $15 per report. As an alumnus of these universities, the cost is $20 per report. For more information call 1-800-926-4JOB. Visit the career center at your university or college to see if it belongs to Information Partners. For information on becoming an Information Partners member, call 415-826-1750.

The National Business Employment Weekly

The National Business Employment Weekly, published by Dow Jones & Company, Inc., is the nation's preeminent career guidance and job-search publication. It offers all regional recruitment advertising from its parent publication, *The Wall Street Journal*, as well as timely editorials on how to find a new job, manage the one you have, or start a business. You will find information on a wide range of careers. You will also get the latest on business and franchising opportunities, and special reports on workplace diversity. To view additional NBEW articles, subscription information, and job hunters' résumés, go to **http://www.occ.com/occ**

CHAPTER 8
Using Cyberspace To Find a Job

In this chapter, you will learn how to

- ❧ find companies with job opportunities;
- ❧ use the Internet as a tool for learning about job resources;
- ❧ develop on-line résumés to showcase talent and skill;
- ❧ find Internet sites to post your résumé with;
- ❧ use the Internet as a tool to maximize your potential for finding a job; and
- ❧ prepare for a job interview by researching prospective companies.

The Internet provides new opportunities for job-seekers and companies to find good employment matches. Many companies are turning to the Internet believing that the people who keep up with the most current information and technological advances in their field are the best candidates for positions. The growing perception among employers is that they may be able to find better candidates if they search on-line.

The types of jobs offered on the Internet have changed dramatically over the last ten years. In the past, job announcements were primarily academic or in the field of science and technology. Now, thousands of positions in all fields from graphic artists to business and marketing professionals, from medical professionals to Internet surfers and Web programmers, are being advertised.

Many companies realize the impact of the digital revolution on business and are searching for professionals who are already on-line cybersurfing, networking with peers, researching information, asking questions, and learning collaboratively from others around the world. A number of companies report difficulty finding such qualified individuals.

How Can the Internet Help Me Find a Job?

The Internet provides an abundance of job resources including searchable databases, résumé postings and advertising, career planning information, and job-search strategies. There are several databases and newsgroups that allow you to post your résumé at no cost. Many companies post job listings on their Web pages.

The Internet also encourages networking with people around the country and around the world. People that you meet on the Internet can be important resources for helping you to find a job and learn more about the business or career you are interested in.

Each day, the number of job openings increases as new services become available. Many believe that the real changes and opportunities are still to come. The question now is no longer whether the Internet should be used to find a job or an employee, but rather, how to use it.

How Do I Begin?

Listed below are a few ways to use the Internet in your job search:

- Research companies that you are interested in by finding and exploring their Web pages. For example, Intel's Home Page has links to job openings listed by geographic location, function, and level. **http://www.intel.com**

- Learn more about job resources, electronic résumés, and employment opportunities available on the Internet.

- Create an electronic résumé.

- Use the Internet to give yourself and your résumé maximum visibility.

- Participate in Usenet newsgroups and listserv mailing lists to network and learn about companies you are interested in working for.

- Learn as much as possible about a prospective company before going for a job interview.

Seven Steps To Internet Job Searching

STEP 1
Research companies that you are interested in by finding and exploring their Web pages. There are many ways to find companies to match your personal and professional needs. Use the information from your self-assessment to refine and define your search for companies. Use both on-line and off-line resources. Listed below are sources to assist you with finding companies.

- Go to your library and review publications in your field of study. Look for classified ads in these publications. Find names of companies that interest you. Research these companies using the search tools you learned about in Chapter 5.

- Search the classified section in newspapers in the cities or regions where you would like to live. Use the Internet to research these companies.

- Use Internet search tools described in Chapter 5 for finding companies and employment opportunities. Begin by using broad terms such as *employment* or *employment and manufacturing*. If you are looking for electronics companies, you might enter in a keyword such as *electronics, electronics companies, manufacturing companies, semiconductors*—or if you know the name of the company, do a search entering the company name as your keyword.

Visit these Web sites to find electronic or manufacturing companies.

http://www.ctrl-c.liu.se/other/admittansen/netinfo.html

http://www.scescape.com/WorldLibrary/business/companies/ elec.html

Another Internet resource for finding employement information is the World Wide Web Virtual Library. **http://www.w3.org/hypertext/ DataSources/bySubject/Overview.html**

STEP 2
Explore job resources and employment opportunities available on the Internet. Many Web sites have job postings and information on how to write résumés and effectively use the Internet for finding a job. Listed below are several excellent Internet resources to help you begin.

- **Employment Opportunities and Job Resources on the Internet** Margaret F. Riley's Web site has excellent job resources. A MUST VISIT Internet stop.
http://www.jobtrak.com/jobguide

- **JobHunt: A Meta-list of On-line Job-Search Resources and Services**
http://rescomp.stanford.edu/jobs.html

- **Job Search and Employment Opportunities: Best Bets from the Net**, Phil Ray and Brad Taylor, University of Michigan SILS
http://asa.ugl.lib.umich.edu/chdocs/employment

- **Job Search Guide**
gopher://una.hh.lib.umich.edu/00/inetdirsstacks/employment %3araytay

- **RPI Career Resources**   **http://www.rpi.edu/dept/cdc**

Usenet newsgroups and listserv mailing lists are two other Internet resources for learning about employment opportunities and for finding out how to find the information you are searching for.

STEP 3

Learn about electronic résumés. The World Wide Web has created opportunities for new types of résumés and business cards. Those individuals who take advantage of the power of this new medium stand out as being technologically advanced and in touch with the future.

Listed below are Web sites to visit to examine on-line résumés. The individuals who have created these résumés understand how to use the medium to sell themselves. At the same time, they are stating that they have special skills for this new marketing medium that sets them apart from other candidates.

Visit these Web sites and study these on-line résumés. Ask yourself the following questions as you look at them:

- How are on-line résumés different from traditional résumés?

- How do on-line résumés have an advantage over traditional résumés?

- What are some characteristics of the Internet as a medium that can be used to your advantage when designing a résumé to sell yourself to a company?

- How are these individuals taking advantage of the Internet as a medium to communicate?

- What do you view as advantages of using on-line résumés?

Sample On-line Résumés and Home Pages

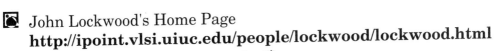 John Lockwood's Home Page
http://ipoint.vlsi.uiuc.edu/people/lockwood/lockwood.html

Mike Swartzbeck Home Page
http://myhouse.com/mikesite/

Sandra L. Daine—Web designer, author, editor
http://q.continuum.net/~shazara/resume.html

Jon Keegan—Illustrator
http://web.syr.edu/~jmkeegan/resume.html

Allan Trautman—Puppeteer and actor
http://www.smartlink.net/~trautman/

Ricardo Araiza—Student
http://pwa.acusd.edu/~ricardo/resume.html

Kenneth Morril—Web Developer
http://webdesk.com/resumes/kjmresume.html

Laura Ann Wallace—Attorney
http://lagnaf.isdn.mcs.net/~laura/

Visit CareerMosaic to learn more about what on-line résumés look like and what they will look like in the future.
http://www.careermosaic.com/

Then go to the Career Resource Center. Send e-mail to these individuals and see if they have received job offers. Ask them about the responses they have received from their Web sites. How did they get maximum exposure to the Internet business community?

STEP 4

Visit on-line sites for job seekers. The next step is to visit Web sites that post résumés. Identify sites where you would like to post your résumé. There are many services available for job-seekers and for companies looking for employees. Companies usually pay to be listed; job-seekers may be allowed to post their résumés at no cost.

America's Job Bank
This on-line employment service offers information on over 250,000 employment opportunities. **http://www.ajb.dni.us/index.html**

CareerMosaic
Begin your CareerMosaic tour by visiting the J.O.B.S. database, with thousands of up-to-date opportunities from hundreds of employers. Then stop by their USENET "jobs.offered" page to perform a full-text search of jobs listed in regional and occupational newsgroups in the U.S. and abroad. If you would also like to make your résumé accessible to interested employers from all corners of the globe, key into ResumeCM and post your résumé on-line.
http://www.careermosaic.com

Career Path
Review employment opportunities from a number of the nation's leading daily newspapers such as *The New York Times*, *Los Angeles Times*, *The Boston Globe*, *Chicago Tribune*, *San Jose Mercury News*, and *The Washington Post*. **http://www.careerpath.com**

Career Resources Home Page
This Web site has links to on-line employment services including professional and university-based services.
http://www.rpi.edu/dept/cdc/homepage.html

CareerWeb
Search by job, location, employment, or keyword to find the perfect job. You can also browse employer profiles and search the Library's list of related publications. **http://www.cweb.com**

E-Span
E-Span, one of the countries foremost on-line recruitment services, provides tools designed to meet the needs of an increasingly competitive career market. Recently added to their services is Résumé ProKeyword Database that is available to more than 60,000 individually registered career service consumers. Visit this Web site and select Job Tools. **http://www.espan.com**

Helpwanted.com
This site offers a searchable index of job openings for companies that have paid to be listed. **http://helpwanted.com**

IntelliMatch
Connect to IntelliMatch and fill out a résumé; hundreds of employers will have access to your profile via the Holmes search software. Review other services such as job-related sites and products, participating companies, and descriptions of available jobs.
http://www.intellimatch.com

The Internet Online Career Center
This career center and employment database is one of the highest-volume job centers with a long list of employment opportunities and resources. Post your résumé in HTML format. Use multimedia (images, photographs, audio, and video) to enrich your résumé.
http://www.occ.com

The Monster Board
This unusual ad agency is a service for recruitment and furnishes information for job-seekers. **http://www.monster.com/home.html**

Stanford University
Stanford University's site provides listings of on-line job services such as Medsearch and the Chronicle of Higher Education. They also have links to other agencies. **http://rescomp.stanford.edu**

STEP 5
Create an on-line résumé to showcase your talents. In Steps 1-4, you learned about

- companies that fit your career interests;
- job and career resources on the Internet;
- electronic résumés and how they can showcase talents; and
- Web sites for job seekers.

You are now ready to use this information to create your own electronic résumé to showcase your talents and skills. Well designed, creative, and interesting online résumés set interested job seekers apart from others. When many individuals are competing for the same job, it is essential to stand apart from others and showcase your talents as to how they will benefit a company, especially in a time when businesses realize the importance of being networked to the world.

Creating an exceptional on-line résumé takes planning and careful thought. On-line résumés take different form. Some may be an electronic version of a text-based résumé. Others may be home pages with links to resources that showcase a person's work and expertise.

Preparing Your Résumé for the Internet

Before you begin, think about your goals and what you would like to accomplish with an on-line résumé. Your primary goal is hopefully to find a job and not just to impress friends with a cool Home Page.

You will need to determine whether to create your on-line résumé yourself or hire a résumé service. If you are creating an electronic résumé on your own, consider whether you want to develop your own Home Page for your résumé or use an on-line database service to post your résumé. If you plan to create a Home Page you will need to learn HTML programming or use a software application program that creates an HTML code from your text. There are many software programs to assist you with this, as well as word processing programs that convert text to HTML.

If you are using an on-line résumé service, find out what type of text file they want. Usually, you will be asked for ASCII text. Most word processors and résumé writing programs have options for saving a file as ASCII or plaintext.

Investigate Résumé Services

Consider using a résumé service to create an on-line résumé. One advantage of using a service is that you may be able to get your résumé on-line quickly with instant exposure to many job opportunities. One disadvantage of using a service is that you do not have as much control over how your résumé will look. You may not be able to use complex graphics or other multimedia effects when using a service.

Cost is another limiting factor. Some companies charge a monthly fee to post your résumé, in addition to a set-up and sign-on fee. Look for companies that charge a reasonable fee to write a résumé ($35 - $50) and no fee to post it on their Web site. Investigate what other services they provide. How many visitors does this site have each day, each week? Will this site give you maximum exposure to potential employers?

Visit Web sites with résumé services. Evaluate their services: How many online résumés are posted? Are the résumés well done, creative, interesting? How well do they promote the job seeker?

Whether you choose to use a résumé service or to create your own on-line résumé, there are seven essential elements to follow.

The Seven Essential Elements of Electronic Résumés

1. **Text must be properly formatted as an ASCII text file.**
   Using ASCII ensures that your résumé can be read universally by everyone and that readers will be able to scroll through your text. Additionally, an ASCII document can be e-mailed to anyone in the world and read.

2. **Showcase your experience and education.**
   At the top of your résumé, provide links to your experience and education. Experience is usually the first thing employers look for. A fancy résumé will not help you get a job if you do not have the right qualifications. Notice that many on-line résumés provide examples of their work.

3. Provide an e-mail hyperlink.

An e-mail hyperlink provides an easy way for prospective employers to contact you easily by e-mail. By clicking on the link, they can send you a message, ask questions, or request additional information. Anything that makes it easier for recruiters improves your chances of being called for an interview.

4. Use nouns as keywords to describe your experience.

When employers use the Internet to search for qualified individuals, they will frequently use search engines that require keywords. The keywords used by employers are descriptors of the essential characteristics required to do a job, such as: education, experience, skills, knowledge, and abilities. The more keywords that your résumé contains, the better your chances are of being found in an electronic database.

Action words such as *created*, *arbitrated*, *managed*, *designed*, and *administered* are out. Therefore, use words such as *manager*, *electronics engineer*, *accountant*, *MBA*. The use of nouns will tend to produce better results.

5. Use white space.

An electronic résumé does not need to be one page long and single spaced. The use of white space makes reading easier and is visually more appealing. Use space to indicate that one topic has ended and another begun.

If you are a new graduate, a résumé equivalent of one page is appropriate. For most individuals with experience, the equivalent of two pages is the norm. Individuals who have worked in a field for many years may use two to three pages.

6. Keep track of the number of visitors to your page.

A counter will keep track of the number of visitors that view your page. A counter is important when paying for a résumé service to monitor how successful the service is with getting exposure for your résumé.

7. Be sure your page gets maximum exposure to potential employers.

One way that employers look for prospective employees is to do a keyword search using search engines such as Yahoo, Excite, and Alta Vista. Each search engine uses a different criteria for selection of Internet resources that are available in their database. Be sure that your page is listed with search engines. Visit search engine Web sites and learn how to submit your page. Additionally, investigate the selection criteria for these search engines.

■ ■

Other Points To Consider

- Think of creative ways to show your talents, abilities, and skills. World Wide Web pages are excellent for linking to examples of your work.

- REMEMBER that experience is perhaps the critical element for recruiters. Be sure your résumé showcases your experience and skills in as many ways as possible.

- Visit the top 10-15 companies that you are interested in working for. Research their World Wide Web Home Page. Learn as much as possible about the company. Use this information when designing and creating your résumé to include information and skills that the company is looking for in their employees. Use this information before you go for an interview to show your knowledge and interest in the company.

- Investigate whether you will be able to submit your résumé electronically to the company.

- Are you concerned about confidentiality? Inquire about who will have access to the database you are posting your résumé with. Will you be notified if your résumé is forwarded to an employer? If the answers to these questions are not satisfactory to you, reconsider posting with this database.

- Once you post your résumé, anyone can look at it and find your address and phone number. You may want to omit your home address and just list your phone number and an e-mail hyperlink. Many recruiters and employers prefer to contact individuals by phone; if you decide not to post your phone number, you may be overlooked.

- Can your résumé be updated at no cost? You may want to add something to your résumé or correct a typo. Look for services that do not charge for updates.

- How long will your résumé be posted with the service? A good service will delete résumés after 3 to 6 months if they have not been updated.

STEP 6

Use the Internet to give yourself and your résumé maximum visibility. Successful job searches using the Internet require an aggressive approach. A résumé should be filed with many job-listing databases as well as with companies that you are interested in working for. Listed below are additional guidelines for giving yourself maximum exposure using the Internet.

- File your résumé with as many databases as possible. Visit the Net Sites for Job Seekers and find as many sites as possible to submit your résumé to.

- Use search engines and their indexes to locate resources specific to your occupation of interest.

- Visit the Home Pages of companies that you are interested in and explore their pages to find job listings. Find out if you can submit your résumé to them electronically.

- Use Usenet newsgroups and listserv mailing lists for information on finding jobs and posting your résumé.

Listed below are several Usenet newsgroups to investigate for posting résumés.

biz.jobs.offered
misc.jobs.offered
misc.jobs.offered.entry
misc.jobs.contract

STEP 7

Learn as much as possible about a prospective company before going for a job interview. Before going for a job interview, it is important to learn as much as possible about the prospective company. The Internet is an excellent tool to assist you with finding up-to-date information about a company. Annual reports and information found in journals, books, or in the library will not be as current as what you will find on the Internet. World Wide Web sites are constantly being changed and updated.

The information that you should be investigating about a prospective company includes:

- What are the company's products and services?
- Who are the company's customers?
- What is the size of the company? Has the company grown over the last five years?
- Is the company profitable?
- Has the company laid off employees?
- How do customers and competitors view the company's products and services?
- Who are the company's major competitors?
- What is the corporate culture like?
- Is the employee turnover rate low, high, or average?
- Are work schedules flexible?
- How many hours a day do employees work?
- What is the typical hiring process?
- Is the organization non-profit or for-profit? There are differences in how these types of organizations operate.

CHAPTER 9
Guided Tour...
Using the Internet For
Career Exploration and
Job Opportunities In Electronics

● ●

Graduates from electronics programs will more than likely be searching for jobs in one of these two design areas: electronic circuits and semiconductor circuits.

This chapter provides a guided tour of

- ◆ how to use the Internet for career exploration;
- ◆ how to use the Internet to find job opportunities in the field of electronics;
- ◆ World Wide Web sites to explore for employment and job opportunities;
- ◆ Internet resources for finding information on companies; and
- ◆ Internet Business Directories.

Job Search 1 uses Internet resources for finding electronic circuit design jobs. You will also learn how to research companies with posted jobs to learn more about their corporate culture and work environment.

Job Search 2 uses different Internet resources for finding information on semiconductor circuit design jobs.

● ●

Job Search 1

Finding Employment Opportunities—Electronic Circuit Design

There are many excellent Internet resources for assisting you with your career exploration. This guided tour takes you on a journey to Web sites with career and business resources for employment opportunities in electronics, as well as information about electronics and manufacturing companies.

> ### NOTE
> There are many ways to find employment opportunities and information about businesses. The more knowledgeable you are about using Internet resources for finding information, the more options you will have open to you. This example merely serves as one pathway you might take.

STEP 1

Explore job opportunities. Visit Internet's Online Career Center to search for electronic circuit design jobs.

 http://www.occ.com/occ

FIGURE 9.1

Online Career Center's Home Page used to find job opportunities

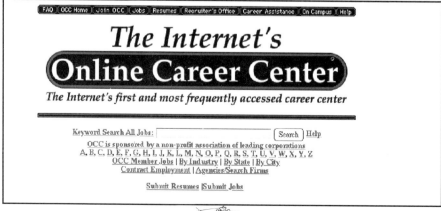

This resource has links to Frequently Asked Questions, Jobs, Résumés, Career Fairs and Events, Career Assistance, and much more. Before you begin, you may want to explore some of these links. Visit the Career Assistance Center for information on writing electronic résumés and how to submit a résumé to the Online Career Center.

STEP 2

Search for a job. Use the search tool at this site to first search for design of electronic circuits jobs. In the **Keyword Search** field, type in the job description; in this case, *designing electronic circuits*. Click on the **Search** button. See Figure 9.2 for the results of this search.

FIGURE 9.2

Search results from the Online Career Center using keywords, *designing electronic circuits*

---

1. [Mar 27] US-PA - ELECTRONIC ENGINEER - HEC GROUP
2. [Mar 25] US-FL-ELECTRICAL ENGINEER(s)-WESTINGHOUSE POWER GENERATION
3. [Mar 19] US-CA-San Jose - Circuit Design Engineer - Fujitsu
4. [Mar 11] ENGINEERING OPPORTUNITIES--Siemens Medical
5. [Feb 22] Electronics EngineerChandler Evans Control Systems Division

---

When you click on the first link to <u>ELECTRONIC ENGINEER - HEC GROUP</u>, you are given this information about the job.

```
The ideal candidate will be a BSEE or equivalent with 5+ years
experience in embedded circuit design. Will be involved in designing
of hardware circuits for use in computerized systems and test
equipment, performing pcb layouts, designing test procedures,
coordinating project activities, writing documentation etc.

$: 40k - 55k

For further information: HEC Group: 1-905-648-0013
Fax: 1-905-648-7016; E-mail: HEC@hec-group.com
Other positions available across North America:
    http://www.hec-group.com/~hec/

xhecx
```

FIGURE 9.3
Web page for HEC Group
http://www.hec-group.com/~hec/

The HEC Group lists a Web site below their job description. Write down the URL and visit its site for more information on the company.

STEP 3
Note companies of interest. In this example, HEC Group is noted.

STEP 4
Learn about companies where jobs are offered. Your search query provided the names of several jobs, including Siemens Medical. To learn more about Siemens Medical, research the company.

FIGURE 9.4

Home Page for Siemens

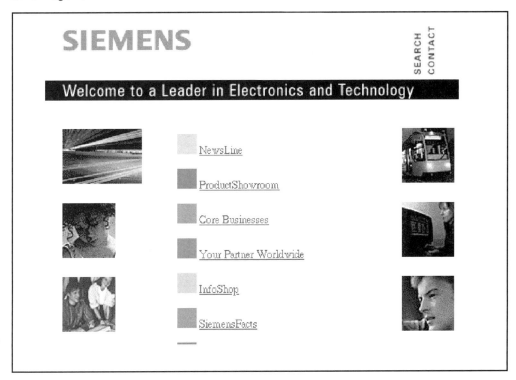

There are many ways to learn more about companies. In this instance, visit one Web site for electronic resources and find links to Siemens Components (see Fig. 9.4).

http://www.ctrl-c.liu.se/other/admittansen/netinfo.html

After clicking on the link to Siemens Components, view and explore their Home Page to learn more about the company, its products, and services.

Job Search 2

Finding Employment Opportunities—Semiconductor Circuit Design

As you explore and use career and job-related Internet resources, you will find that there is no single Web site that will provide all the information and tools that you will need to help you find a job.

You will find that many Web sites have a search tool to help you locate jobs and companies. Before you conduct a search, be sure you understand how the search engine can be used most effectively to find information. Select links to **Options** or **Help** to learn more about the search tool. Other pages will provide you with information on how to conduct a search. For example, in Job Search 1, you visited the Online Career Center and found five electronics jobs by entering the keywords, *design of electronic circuits.*

The search engines at other Web sites may use different criteria. In Job Search 2, you will use another search tool that requires different types of keywords to be entered to find the same type of information on job opportunities.

STEP 1
Explore jobs opportunities. Visit America's Job Bank to search for semiconductor circuit design jobs. **http://www.ajb.dni.us/index.html**

FIGURE 9.5
Home Page for America's Job Bank

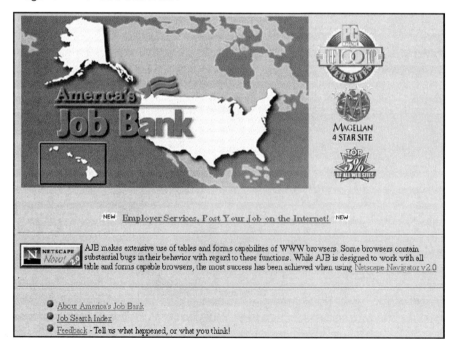

STEP2
Select the link to the Job Search Index (shown in Fig 9.6).

FIGURE 9.6
Clicking on the **Job Search Index** link, takes you to the Job Search Index Web page

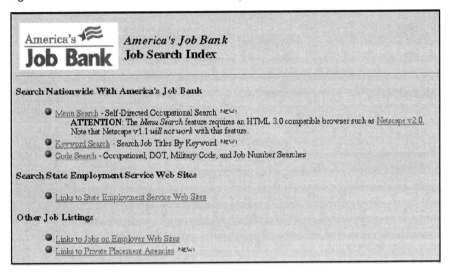

FIGURE 9.7
America's Job Bank page for finding a job by location and job title

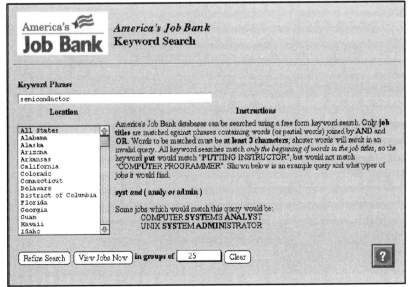

STEP 3

Select keyword search by job titles. Notice the difference in searching criteria used by the America's Job Bank in comparison to the Online Career Center. To search for a job, we must enter a job title. If we entered *design of semiconductor circuits*, the search engine would not produce any results.

Read the following instructions for conducting a search

> America's Job Bank databases can be searched using a free form keyword search. Only job titles are matched against phrases containing words (or partial words) joined by AND and OR. Words to be matched must be at least 3 characters; shorter words will result in an invalid query. Listed below are examples of job titles.
>
> COMPUTER SYSTEM'S ANALYST
> UNIX SYSTEM ADMINISTRATOR

Using the American Job Bank search engine, we must identify the job title for the design of semiconductor circuits. Several job descriptions would include:

- semiconductor architecture
- integrated circuit design
- VLSI engineer

After entering in these keywords for three different searches, the Job Bank indicates no matches found. The next option is to enter in just the keyword *semiconductor*. This keyword search yielded two job postings.

122

FIGURE 9.8

Search results using America's Job Bank and entering the keyword *semiconductor*

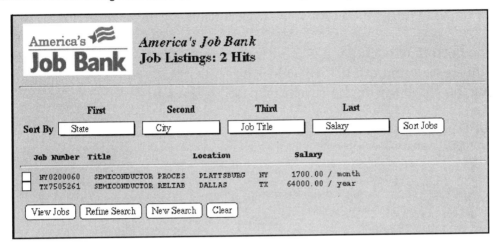

Notice that this search engine will sort jobs in an order that you select: city, state, job title, or salary. For this search there were only two jobs, therefore sorting was not necessary.

To learn about each job, click once in the box next to the job posting. Then click on the button, **View Jobs**.

FIGURE 9.9

Job description for a semi-conductor job

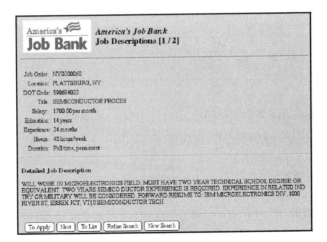

Notice there is no company name given with this job description.

Click on the **Next** button for a job description of the second job.

At the bottom of each page is a button to apply for the job if you feel you are interested and qualify. By selecting this option, you may learn more about the company or the recruiter posting the job.

FIGURE 9.10
An example of an on-line application form for a job

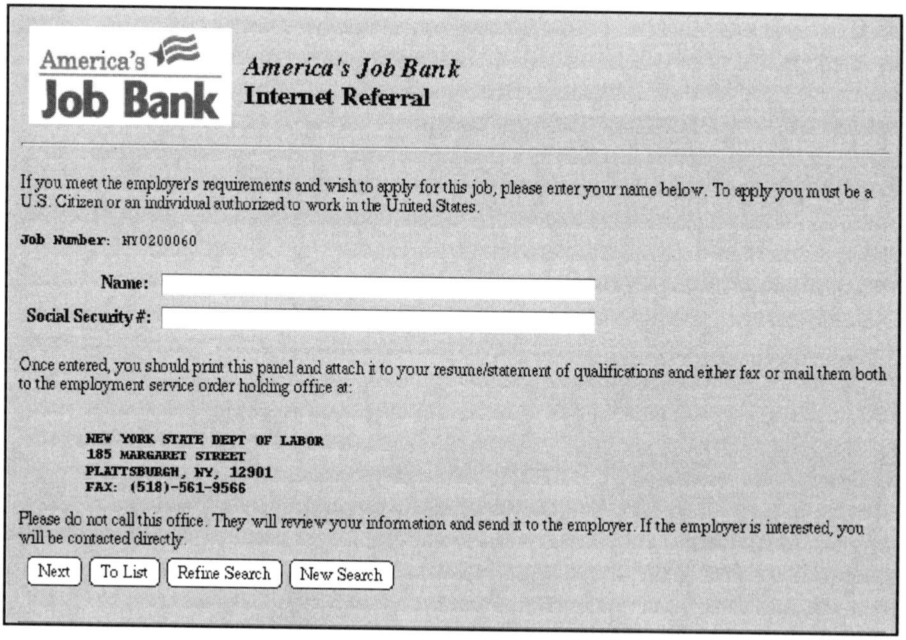

America's Job Bank
America's Job Bank
Internet Referral

If you meet the employer's requirements and wish to apply for this job, please enter your name below. To apply you must be a U.S. Citizen or an individual authorized to work in the United States.

Job Number: NY0200060

Name:

Social Security #:

Once entered, you should print this panel and attach it to your resume/statement of qualifications and either fax or mail them both to the employment service order holding office at:

NEW YORK STATE DEPT OF LABOR
185 MARGARET STREET
PLATTSBURGH, NY, 12901
FAX: (518)-561-9566

Please do not call this office. They will review your information and send it to the employer. If the employer is interested, you will be contacted directly.

Next | To List | Refine Search | New Search

When the **Apply** button for both of these jobs is selected, you see in Figure 9.10 that they are being handled by an employment service.

STEP 4

Learn more about the cities where these jobs are located. After reading the job description you may be interested in applying for the job, but would first like to learn more about the city where the job is located.

124

To learn about the cities for these two jobs, visit City Net.
http://www.city.net

This Web site (Fig. 9.11) provides options for learning about cities by visiting their designated Most Popular Cities or by conducting a keyword search.

Since one of the semiconductor jobs was in Dallas, select the link to Dallas (Fig. 9.12)

FIGURE 9.11
Web site for City Net

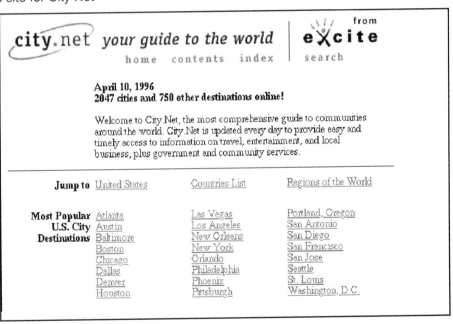

FIGURE 9.12

City Net's link to information on Dallas

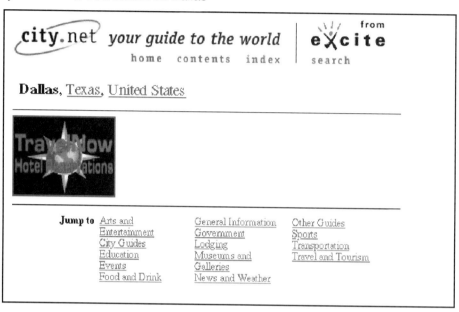

City Net takes us to links where we can explore and learn more about Dallas.

The second city with a semiconductor job posting was Plattsburg, New York. To find information on this city, we will use City Net's search tool and enter the keywords *Plattsburg, New York*.

FIGURE 9.13

Search results for City Net's search on Plattsburg, New York

81% ITVA Region 2 - North Atlantic
Summary: Region 2 Vice-President: George Cauttero Gcautter@aol. North Jersey - Hotline: (201) 802-4868.

80% BNL METEOROLOGY
Summary: New York City, NY State zone , NY Coastal, Bridgeport, the New York State discussion, simplifie discussion and the NGM Model Output Statistics (MOS) for Islip. --> Worth checking is NOAA/NWS Forec Office - NEW YORK, NY ! BEWARE: Products obtained from the INTERNET may not be timely!

79% Adirondack Mountain Club
Summary: Lake Placid - Lake Placid. New York Metropolitan New York City.

79% Cen-Com Internet - Site Locations
Summary: Listed here are locations around the country where Cen-Com Internet, Inc. Find out more about Cen-Com Internet by e-mail at info@cencom.

STEP 5

Use other career and job resources to search for jobs. Two other excellent Internet sites to visit to assist with finding jobs are: Career Web and the Monster Board. Their job searches differ from the Online Career Center and America's Job Bank. Enter in this URL for Career Web. **http://www.cweb.com**

FIGURE 9.14
Career Web's search tool for finding a job by discipline and state/country

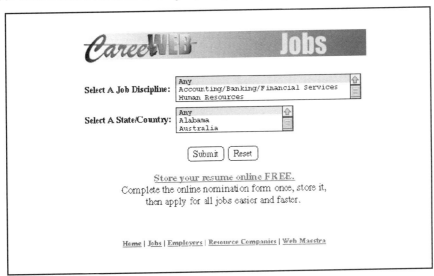

Notice that this Job site asks you to select a job discipline and a state or country of your preference. Scroll down through the job disciplines to find one that best matches the job you are searching for. For your first search you may want to select *Any State/Country* to see what is available. Later you can refine your search.

STEP 6

Visit the Monster Board to investigate its job search resources. **http://www.monster.com/home.html** is its URL. When you connect to the Monster Board, sign in as a first-time visitor. There is no charge for using and visiting this site.

After clicking on the two links, job opportunities and career search, we are at the page shown in Figure 9.15.

This job site presents job locations and job disciplines for you to select. For more information on each job discipline, click on the link for more information. After you select a location and discipline and conduct a search, you maybe asked for refine your search.

> ### NOTE
>
> After using four different Internet Job resources, you now have some knowledge about Internet tools for finding job opportunities. You should also have a better understanding of how these resources differ in helping you to find a job. As with all Internet resources, you will need to use many different tools to help you find the information you are searching for. You will find that some tools are more helpful than others for specific types of information requested.

FIGURE 9.15

Monster Board's search options using location and job discipline criteria

Select job locations

For more information on the job locations, select here.

```
-ANY-
AK-ANCHORAGE
AK-FAIRBANKS
AL-ANNISTON
AL-BIRMINGHAM
```

Select job disciplines

For more information on job disciplines, select here.

```
-ANY-
Bio-Chemistry
Bio-Clinical Research
Bio-Engineer
Bio-Environmental Science
```

[Start Search] [Clear Selections]

Internet Resources To Explore

Employment and Job Opportunities

Listed below are some of the best Internet resources to explore to help you find career and job opportunities.

America's Job Bank

This Web site is highly recommended as an excellent resource for employment opportunities. Search for a job by entering a keyword. When the search results are returned, sort your search by city, state, job title, or salary.

The America's Job Bank computerized network links 1800 state employment service offices. It provides job-seekers with the largest pool of active job opportunities available anywhere. For employers it provides rapid, national exposure for job openings. The nationwide listings in America's Job Bank contain information on approximately 250,000 jobs.

In addition to the Internet, America's Job Bank is available on computer systems in public libraries, colleges and universities, high schools, shopping malls, and other places of public access. AJB is also available at transition offices on military bases worldwide.
http://www.ajb.dni.us/index.html

CareerMosaic

This site is an excellent place to begin job hunting. Here you will find listed thousands of job opportunities in every industry and discipline. Many companies are using this site to list their openings. Search this huge database by entering job description, job title, city, state, and country qualifiers to tailor your job listing to specific requirements.

Also available is CareerMosaic Usenet Search Page with thousands of new job postings each day.
http://www.careermosaic.com/cm/cm1.html

Career Path

This MUST VISIT Web site provides excellent search options for a job-by-job category matched with listings in national newspapers: *The Boston Globe, Chicago Tribune, Los Angeles Times, The New York Times, San Jose Mercury News,* and *The Washington Post.*
http://www.careerpath.com

Career Web

Career Web provides links to employers, a Resource Center, newsletter, and a search engine for finding jobs.
http://www.cweb.com

E•Span

Visit E•Span to search for jobs by keywords, company name, geographic location, educational level, years of experience, job level, or salary desired. **http://www.espan.com**

Monster Board

The Monster Board is a commercial site on the Web operated by ADION, Inc., a large recruitment advertising agency in New England. This very graphic and colorful site has a browse and select option for searching for a company by name, location, discipline, or specific job title. Most of the job postings are technical and located on the East Coast. When you find a job you are interested in, use the on-line form to send in your application or e-mail in your résumé.
http://www.monster.com/home.html

The National Business Employment Weekly

The National Business Employment Weekly, published by Dow Jones & Company, Inc., is the nation's preeminent career guidance and job-search publication. It offers all regional recruitment advertising from its parent publication, *The Wall Street Journal*, as well as timely editorials on how to find a new job, manage the one you have, or start a business. You'll find information on a wide range of careers. You'll also get the latest on business and franchising opportunities, and special reports on workplace diversity. To view additional NBEW articles, subscription information, and job hunters' résumés, go to
http://www.occ.com/occ

NetJobs
This Web site has links to career listings, on-line résumés, a consultant's corner, employment information, a net job's e-mail list, and much more.
http://www.netjobs.com:8000/

Finding Information on Companies

There are many other Internet resources for finding information on companies. Resources you will want to investigate include:

- Internet sites with listings of electronic and manufacturing companies

- Web sites with links to electronic resources

- Links to or a URL for a company's Web site

- Internet Business Directories

- Internet Navigational Directories

- Search engines

Internet Links To Electronic and Manufacturing Companies

Listed below are several URLs that have listings of electronic and manufacturing companies:

http://www.ctrl-c.liu.se/other/admittansen/netinfo.html

http://www.scescape.com/WorldLibrary/business/companies/elec.html

Links To Electronic Resources

The EE/CS Mother Site is sponsored by the Stanford IEEE with more than 250 companies listed. This site is an excellent starting point when searching for a job or when looking for Electrical Engineering and Computer Science related information.
http://www-ee.stanford.edu/soe/ieee/eesites.html

Internet Business Directories

Apollo: This Web site provides options for searching for a company by country and keyword. **http://apollo.co.uk**

Linkstar
Linkstar provides keyword search options, plus a listing of categories for finding information. **http://www.linkstar.com**

New Rider's WWW Yellow Pages
Riders' World Wide Web Yellow Pages offers business search options by keyword or by category. A keyword search for jobs produced numerous job related links. **http://www.mcp.com/newriders/wwwyp**

Nynex Interactive Yellow Pages
The largest of the business directories provides options for searching for companies by business location, category, or business name. **http://www.niyp.com**

Virtual Yellow Pages
The Virtual Yellow Pages is a comprehensive and easy-to-use directory of Web sites and information. **http://www.vyp.com**

World Wide Yellow Pages
The World Wide Yellow Page has a link to assist you with finding information on businesses. **http://www.yellow.com**

Refer to Chapter 5 for Internet navigational directories and search engines.

CHAPTER 1O
Learning Adventures...
Career Exploration and Job
Opportunities in Electronics

This chapter provides learning activities for applying and using the information presented in *Internet Investigations in Electronics*. Activities include

- ➼ using an Internet browser to visit Web sites
- ➼ using Bookmarks to save your favorite Internet sites
- ➼ subscribing to a listserv mailing list
- ➼ exploring Usenet newsgroups
- ➼ exploring electronic chats
- ➼ using search tools to find information
- ➼ exploring cool electronic and electrical sites to learn about how the field of Electronic Technology is using the Internet
- ➼ a self-awareness journey to learn about career goals and interests
- ➼ using the Internet as a tool for career exploration
- ➼ using the Internet to find job opportunities
- ➼ learning about electronic résumés
- ➼ designing an electronic résumé
- ➼ using the Internet to research companies you are interested in working for

Chapter 1: What Is the Internet?
The metaphorical Internet—traveling down the information superhighway

1. Write a metaphor for the Internet.

2. Today there are between 30-50 million Internet users. How do you see the Internet impacting your life? Your career?

3. Being on the Internet means having full access to all Internet services: electronic mail, Telnet, File Transfer Protocol (FTP), and the World Wide Web. Make a map that illustrates your understanding of what it means to be connected to the Internet.

Chapter 2: Guided Tour—Internet Browsers
I'm not lost . . . I'm just exploring!

Visit these Web sites using either Netscape Navigator or Microsoft's Explorer.
http://www.cybertown.com
http://www.microsoft.com
http://www.kbt.com/gc
http://www.paris.org
http://www.timeinc.com/pathfinder/Greet.html
http://mosaic.larc.nasa.gov/nasaonline/nasaonline.html
http://espnet.sportzone.com

1. Make bookmarks of Web pages you would like to save.

2. Organize your bookmarks by categories by creating file folders.

3. How do you think that the Internet will be used in the future for communication? For business? For personal use? In the field of electronics?

4. Design a personal Home Page that provides information about yourself.

5. How is the Internet being used for communication?

6. Is the Internet a useful tool or just a fun, new technology that produces the "Oh Wow . . . this is cool" experience?

7. Do you believe that the Internet has a role in the future for your career and professional life? If so, what?

8. What do you believe is the best application for the Internet?

9. Export your bookmarks to a floppy disk.

10. Import your bookmarks from a floppy disk to the computer you are using.

Chapter 4: Chatting on the Net
The Internet opens new doors to virtual communities where we step through the looking glass

1. Visit this Web site and search for a listserv in your field of study. **http://www.tile.net/tile/listserv/index.html**

2. Subscribe to several listserv mailing lists.

3. Use Netscape to explore Usenet newsgroups. Look for 5-10 groups related to your field of study.

4. Visit these Web sites to experience Internet chat.
Time Warner's Pathfinder **http://www.pathfinder.com**
HotWired **http://www.hotwired.com**
The Palace **http:www.thepalace.comg**
Globe **http://globe1.csuglab.cornell.edu/global/homepage.html**

5. Do you see a use for on-line chats for business in the future?

Chapter 5: Finding Information and Resources
Traveling the Internet without getting caught in the Web

Select a topic of interest (i.e., a hobby, sport, country, or a trip you plan to take).

1. Use Yahoo and Magellan to research your topic. Begin by investigating their subject directories. After you have explored the directories, do a keyword search.

2. How do Yahoo and Magellan differ in the way they provide access to information? Do you find one better or more useful than the other?

3. Explore the Advanced Options in Yahoo to refine and limit your search. Conduct a search using the Advanced Options. Did you get better returns?

4. Use each of the following search engines to research your topic: Excite, Alta Vista, Infoseek, and Open Text. Before you use these search engines, read how to use their advanced options for a more efficient search.

5. Compare and contrast Excite, Alta Vista, Infoseek, and Open Text. Which one did you find most useful in providing the information you were searching for? What are the advantages of each? Disadvantages?

6. How do search tools such as Yahoo and Magellan differ from search engines such as Excite, Alta Vista, Infoseek, and Open Text? Do you think there would be times you would find one more useful than another?

7. Begin a category or a file folder in your Internet browser for Internet Research tools. Make bookmarks of search tools that you find helpful.

8. What does the future hold for Internet search tools?

Chapter 6: Cool Electronic and Electrical Web Sites

The scientist does not study nature because it is useful; he studies it because he delights in it, and he delights in it because it is beautiful. If nature were not beautiful, it would not be worth knowing, and if nature were not worth knowing, life would not be worth living.
—Henri Poincar

1. Journey into the cyberworld of Electronics Engineering. Learn about how the information you are studying in your classes is applied and used to create and build in the real-world of Electronics Engineering.

2. Identify an area of electronics you are interested in. Explore the electronics links to find more information in this area.

3. What companies produce products in your field?

4. Select a technical society or publication and keep up with its development and activities on the Internet.

5. Investigate engineering or trade publications available on the Internet.

6. Do the Ohm's Law experiment.
   http://plabpc.csustan.edu/physics/expt/ohmslaw.htm

7. Use the search tools you learned in Chapter 5 to research topics you are studying in your classes.

8. How are electronics companies using the Internet?

9. What do you believe the future holds for the use of the Internet and electronics?

10. List new applications of electronics that you learned about by exploring electronic and electrical Web sites.

Chapter 7: Using Cyberspace for Career Planning

The privilege of a lifetime is being who you are.
                    —Joseph Campbell

1. Write down 25 things you love to do.

2. What do you do best?

3. What are your greatest achievements?

4. What do you find most rewarding when you work?

5. List and prioritize the 10 things that are most important to you. Which of these could you live without? Which of these are an essential part of your life that you cannot live without?

6. When you were a child, what did you want to be when you grew up?

7. What is your dream job now? Describe the perfect work environment. The perfect boss.

8. What is most important to you when evaluating a job?

9. List 5 jobs that incorporate the things you enjoy doing.

10. List 5 things that you would like to do at work?

11. List 5 ways that you will use people as a resource for learning about jobs and careers.

12. Visit the electronic newsstand and explore electronic publications related to your interests and field of study. **http://www.enews.com/**

13. Research companies that have jobs you are interested in by visiting these two Web sites:

    http://www.ctrl-c.liu.se/other/admittansen/netinfo.html

    http://www.scescape.com/WorldLibrary/business/companies/elec.html

14. Subscribe to a Usenet newsgroup (page 96) and learn more about job opportunities.

15. After you have read postings in a group(s), list 10 things you learned about job opportunities on the Internet.

16. Find a listserv mailing list related to jobs or your field of study (page 97).

17. If you have participated in a Usenet newsgroup or a listserv mailing list, describe your experiences. What have you learned? What are the advantages/disadvantages of a newsgroup? A listserv?

18. Visit job and career resources on the Internet (pages 99-100). Describe information that you found useful.

Chapter 8: Using Cyberspace To Find a Job

To be successful, you must love what you do.
—Dottie Walters

1. Research five companies you are interested in working for. Use the search tools described in Chapter 5. Visit and explore their Web pages. List the products and services of these companies. What have you learned about their work environment? Do they post job openings? How do you apply for a job?

2. Explore job resources and employment opportunities available on the Internet (pages 107-108). List 5 jobs that you find from Internet resources that are of interest to you.

3. Research the city/state where these jobs are located.

4. Learn about electronic résumés by visiting the Web sites with on-line résumés (pages 105-106). How are on-line résumés different from traditional résumés ?

5. What characteristics of the Internet as a medium for communication and information-sharing can be used to showcase your talents and skills with an on-line résumé?

6. What are the advantages/disadvantages of on-line résumés ?

7. Design an on-line résumé that showcases your talents and skills.

8. Visit on-line sites for job seekers (pages 107-108). Explore their on-line résumés. Identify the sites where you would like to post your résumé. What features do you like about each site? What do you have to do to post your résumé? Is there a cost? How long will it be posted? How will you learn if your résumé is seen by prospective employers?

9. You are preparing for an interview with a company that you are interested in working for. You know that it is important to have as much knowledge as possible about this company before the interview. How will you obtain this information? Select one company that you would like to interview with. Research information on this company.

GLOSSARY

applets: Mini applications that a software program such as Netscape downloads and executes.

ASCII (text) files: One of the file transfer modes (binary is another mode) used when transferring files on the Internet. ASCII treats the file as a set of characters that can be read by the computer receiving the ASCII text. ASCII does not recognize text formatting such as boldface, underline, tab stops, or fonts.

binary file: Another transfer mode available for transferring Internet files. In the binary mode, files are transferred which are identical in appearance to the original document.

Binhex (BINary HEXadecimal): A method for converting non-text files (non-ASCII) into ASCII. Used in e-mail programs that can only handle ASCII.

Bit: A single digit number, either a 1 or a zero, that represents the smallest unit of computerized data.

bookmarks: A feature providing the user with the opportunity to mark favorite pages for fast and easy access. Netscape's bookmarks can be organized hierarchically and customized by the user through the Bookmark List dialog box.

boolean operators: Phrases or words such as "and," "or," and "not" that limit a search using Internet search engines.

browser: A client program that interprets and displays HTML documents.

client: A software program assisting in contacting a server somewhere on the Net for information. Examples of client software programs are Gopher, Netscape, Veronica, and Archie. An Archie client runs on a system configured to contact a specific Archie database to query for information.

compression: A process by which a file or a folder is made smaller. The three primary purposes of compression are to save disk space, to save space when doing a backup, and to speed the transmission of a file when transferring over a modem or network.

domain name: The unique name that identifies an Internet site. Names have two or more parts separated by a dot such as **xplora.com**

finger: An Internet software tool for locating people on the Internet. The most common use is to see if an individual has an account at a particular Internet site.

fire wall: A combination of hardware and software that separates a local area network into two parts for security purposes.

frames: A new feature of Netscape Navigator 2.0 that makes it possible to create multiple windows on a Netscape page. Below is an example of a Web page divided into several windows called frames.

FTP (file transfer protocol): Protocol for transferring files between computers on the Internet.

GIF (Graphic Interface Format): A format developed by CompuServe, Inc. for storing complex graphics. This format is one of two used for storing graphics in HTML documents.

Helper Applications: Programs used by Netscape to read files retrieved from the Internet. Different server protocols are used by Netscape to transfer files: HTTP, NNTP, SMTP, and FTP. Each protocol supports different file formats for text, images, video, and sound. When these files are received by Netscape, the external helper applications read, interpret, and display the file.

History List: Netscape keeps track of your Internet journeys. Sites that you visit are listed in the History List found under the **Go** pull-down menu. Click on an Internet site on your list, and you will be linked to that destination.

Home Page: The starting point for World Wide Web exploration. The Home Page contains highlighted words and icons that link to text, graphic, video, and sound files. Home Pages can be developed by anyone: Internet Providers, universities, businesses, and individuals. Netscape allows you to select which Home Page is displayed when you launch the program.

HTML (HyperText Markup Language): A programming language used to create a Web page. This includes the text of the document, its structure, and links to other documents. HTML also includes the programming for accessing and displaying media such as images, video, and sound.

HTTP (HyperText Transfer Protocol): One protocol used by the World Wide Web to transfer information. Web documents begin with **http://**

hyperlinks: Links to other Web information such as a link to another page, an image, a video or sound file.

hypertext: A document containing links to another document. The linked document is displayed by clicking on a highlighted word or icon in the hypertext.

IP address: Every computer on the Internet has a unique IP address. This number consists of four parts separated by dots such as 198.68.32.1

JavaScript: A new programming language developed by Sun Microsystems that makes it possible to incorporate mini-applications called *applets* onto a Web page.

JPEG (Joint Photographic Experts Group): A file format for graphics (photographs, complex images, and video stills) that uses compression.

live objects: Java brings life and interaction to Web pages by making it possible to create live objects. Move your mouse over an image of a house and see the lights go on. Move your mouse to a picture of a woman and hear her welcome you to her Home Page.

MIME (Multimedia Internet Mail Extension): Most multimedia files on the Internet are MIME. The MIME type refers to the type of file: text, HTML, images, video, or sound. When a browser such as Netscape retrieves a file from a server, the server provides the MIME type to establish whether the file format can be read by the software's built-in capabilities or, if not, whether a suitable helper application is available to read the file.

newsgroups: Large distributed bulletin board systems that consist of several thousand specialized discussion groups. Messages are posted to a bulletin board by e-mail for others to read.

NNTP (News Server): A server protocol used by Netscape for transferring Usenet news. Before you can read Usenet news, you must enter the name of your news server to interact with Usenet newsgroups. The news server name is entered in the Mail and News dialog box (**Options** pull-down menu; **Preferences**; Mail and News).

page: A file or document in Netscape that contains hypertext links to multimedia resources.

platform: Netscape Navigator 2.0 is referred to as a platform rather than a browser. A platform program makes it possible for developers to build applications onto it.

PPP (Point-to-Point Protocol): A method by which a computer may use a high speed modem and a standard telephone line to have full Internet access. A PPP or SLIP connection is required to use graphical interfaces with the Internet such as Netscape Navigator and Explorer. Using a PPP or SLIP connection enables you to point and click your way around the Internet.

.sea (self-extracting archives): A file name extension indicating a compression method used by Macintosh computers. Files whose names end in .sea are compressed archives that can be decompressed by double-clicking on the program icon.

search engine: Software programs designed for seeking information on the Internet. Some of these programs search by keyword within a document, title, index, or directory.

server: A computer running software that allows another computer (a client) to communicate with it for information exchange.

shell account: The most basic type of Internet connection. A shell account allows you to dial into the Internet at your provider's site. Your Internet software is run on the computer at that site. On a shell account your Internet interface is text-based. There are no pull-down menus, icons, or graphics. Some Internet providers offer a menu system of Internet options. Others merely provide a Unix system prompt, usually a percent sign or a dollar sign. You must know the commands to enter at the prompt to access the Internet.

SLIP (Serial Line Internet Protocol): A method by which a computer with a high speed modem may connect directly to the Internet through a standard telephone line. A SLIP account is needed to use Netscape. SLIP is currently being replaced with PPP (Point-to-Point Protocol).

SMTP (Simple Mail Transport Protocol): A protocol used by the Internet for electronic mail. Before using Netscape e-mail, the host name of the Internet provider's mail server must be designated. The mail server name is entered in the Mail and News dialog box (**Options** pull-down menu; **Preferences**; Mail and News).

source file: When saved as "source," the document is preserved with its embedded HTML instructions that format the Internet page.

TCP/IP (Transmission Control Protocol/Internet Protocol): The protocol upon which the Internet is based and which supports transmission of data.

toolbar: Navigational buttons used in graphical interface applications.

URL (Uniform Resource Locator): URLs are a standard for locating Internet documents. They use an addressing system for other Internet protocols such as access to Gopher menus, FTP file retrieval, and Usenet newsgroups. The format for a URL is **protocol://server-name:/path**

URL object: Any resource accessible on the World Wide Web: text documents, sound files, movies, and images.

Usenet: Developed in the 1970s for communication among computers at various universities. In the early 1980s, Usenet was being used for electronic discussions on a wide variety of topics and soon became a tool for communication. Today, Usenet groups are analogous to a cafe where people from everywhere in the world gather to discuss and share ideas on topics of common interest.

viewer: Programs needed to display graphics, sound, and video. For example, pictures stored as a GIF image have the file name extension ".gif" and need a gif helper application to display the image. Netscape has the required viewers (external helper applications) built into the software. A list of programs required to view files can be found in the Helper Application menu of Netscape. Open the **Options** pull-down menu, select **Preferences**, then **Helper Applications**.

VRML (Virtual Reality Modeling Language): a programming language that makes 3-dimensional virtual reality experiences possible on Web pages.

REFERENCES

Angell, D. (1996, March). The ins and outs of ISDN. *Internet World*, 78-82.

Bennahum, D. S. (1995, May). Domain street, U.S.A. *NetGuide*, 51-56.

Butler, M. (1994). *How to use the Internet*. Emeryville, CA: Ziff-Davis Press.

Career Center (1996). [On-line]. Available: http://www.monster.com/home.html
    or http://199.94.216.77:80/jobseek/center/cclinks.htm

CareerMosaic Career Resource Center (1996). [On-line]. Available:
    http://www.careermosaic.com/cm/crc/

Conte, R. (1996, May). Guiding lights. *Internet World*, 41-44.

Dixon, P. (1995, May). Jobs on the web. *SKY*, 130-138.

Ellsworth, J. H., & Ellsworth, M.V. (1994). *The Internet business book*. New York:
    John Wiley & Sons, Inc.

Grusky, S. (1996, February). Winning résumé. *Internet World*, 58-68.

Leibs, S. (1995, June). Doing business on the net. *NetGuide*, 48-53.

Leshin, C. (1996). *Internet adventures step-by-step guide to finding and using
    educational resources*. Boston: Allyn and Bacon.

Leshin, C. (1997). *Netscape adventures step-by-step guide to Netscape Navigator
    and the World Wide Web*. New Jersey: Prentice Hall.

Miller, D. (1994, October). The many faces of the Internet. *Internet World*, 34-38.

O'Connell, G. M. (1995, May). A new pitch: Advertising on the World Wide Web
    is a whole new ball game. *Internet World*, 54-56.

Netscape Communication Corporation. (1996, January/February). *Netscape
    Handbook*. [On-line]. Currently available by calling 1-415-528-2555 or
    on-line by selecting the Handbook button from within Netscape.

Reichard, K., & King, N. (1996, June). The Internet phone craze. NetGuide, 52-58.

Resnick, R., & Taylor, D. (1994). *The Internet business*. Indianapolis, IN: Sams Publishing.

Richard, E. (1995, April). Anatomy of the World Wide Web. *Internet World*,   28-30.

Riley, Margaret F. (1996). Employement Opportunities and Job Resources on the Internet [On-line]. Available: http://www.jobtrak.com./jobguide/

Sachs, D., & Stair, H. (1996). *Hands-on Netscape, a tutorial for Windows users*. New Jersey: Prentice Hall.

Sanchez, R. (1994, November/December). Usenet culture. *Internet World*, 38-41.

Schwartz, E. I. (1996, February). Advertising webonomics 101. *Wired*, 74-82.

Signell, K. (1995, March). Upping the ante: The ins and outs of slip/ppp. *Internet World*, 58-60.

Strangelove, M. (1995, May). The walls come down. *Internet World*, 40-44.

Taylor, D. (1994, November/December). Usenet: Past, present, future. *Internet World*, 27-30.

Vendito, G. (1996, June). Internet phones—the future is calling. *Internet World*, 40-52.

Vendito, G. (1996, March). Online services—how does their net access stack-up? *Internet World*, 55-65

Venditto, G. (1996, May). Search engine showdown. *Internet World*, 79-86.

Weiss, A. (1994, December). Gabfest—Internet relay chat. *Internet World*, 58-62.

Welz, G. (1995, May). A tour of ads on-line. *Internet World,* 48-50.

Wiggins, R. W. (1994, March). Files come in flavors. *Internet World,* 52-56.

Wiggins, R. W. (1994, April). Webolution: The evolution of the revolutionary World Wide Web. *Internet World*, 33-38.

Wilson, S. (1995). *World Wide Web design guide*. Indiana: Hayden Books.

Additional Resources

World Wide Web, Gopher, and FTP sites were found using InfoSeek, YAHOO, Lycos, WebCrawler, Alta Vista, VERONICA, and Archie.

APPENDIX I
Connecting To the Internet

....

Connecting To the Internet

There are three ways to connect to the Internet:

- a network
- an on-line service
- a SLIP/PPP connection

Network Connection

Network connections are most often found in colleges, schools, businesses, or government agencies and use dedicated lines to provide fast access to all Internet resources. Special hardware such as routers may be required at the local site. Prices depend on bandwidth and the speed of the connection.

On-line Services

Examples of on-line services include America Online, CompuServe, Prodigy, Delphi, and Microsoft Network. On-line services are virtual communities that provide services to their subscribers including electronic mail, discussion forums on topics of interest, real time chats, business and advertising opportunities, software libraries, stock quotes, on-line newspapers, and Internet resources (Gopher, FTP, newsgroups). There are advantages and disadvantages to these on-line services.

Advantages

The advantages to on-line services include:

- easy to install and use,
- content offered by provider,
- easy to find and download software,
- easy to use e-mail, and
- virtual community of resources and people.

Commercial on-line services are excellent places to begin exploring and learning about the use of e-mail and access to network information and resources.

Disadvantages
The disadvantages to on-line services include:

- expensive to use,
- do not always access all Internet resources such as Gopher, FTP, and Telnet, and
- must use the on-line service's e-mail program and Internet browser.

On-line services charge an average of $9.95 per month for 5 hours of on-line time. Additional on-line time is billed at rates of $2.95 to $5.95 per hour. Some services charge more for being on-line at peak hours such as during the day.

In comparison, an Internet access provider may charge $15-30 per month for 100 to unlimited hours of on-line time. Prices vary depending on your locality and the Internet access provider.

SLIP/PPP Connection
Internet access providers offer SLIP (Serial Line Interface Protocol) or PPP (Point-to-Point Protocol) connections (SLIP/PPP). This service is referred to as *Dial-Up-Networking* and makes it possible for your PC to dial into their server and communicate with other computers on the Internet. Once you have established a PPP, SLIP, or direct Internet connection, you can use any software that speaks the Internet language called TCP/IP. There are several TCP/IP software applications including Eudora, Netscape Navigator, and Explorer.

Internet service providers should give you the required TCP/IP software to get you connected to the Internet. Additionally, many will provide Internet applications such as Eudora, Netscape Navigator, and Explorer. Prices are usually based on hours of usage, bandwidth, and locality.

TCP/IP and SLIP/PPP Software
Macintosh Software

TCP/IP software for the Macintosh is called MacTCP and is supplied by Apple. Two popular software choices necessary to implement either SLIP or PPP are MacPPP or MacSLIP. Using one of these programs with MacTCP creates a direct Internet connection.

Your Internet access provider should give you software that has already been configured for connecting your Mac to the Internet.

Commercial Online Services

America Online
http://www.aol.com
(800) 827-6347
e-mail: postmaster@aol.com

AT&T WorldNet Services
http://www.att.com
(800) 831-5269
e-mail: webmaster@att.com

CompuServe
(800) 848-8990
e-mail: 70006,101@compuserve.com

Microsoft Network
http://www.msn.com
(800) 426-9400

Prodigy
(800) 776-3449

The WELL
http://www.well.com
(415) 332-4335
e-mail: web-info@well.com

APPENDIX II
Finding an Internet Provider

There are several Web sites to help you find an Internet access provider.
http://thelist.com
http://www.clari.net/iap/iapcode.htm
http://www.primus.com/providers

To find the names of providers in your area, click on the link to your area code. You will find descriptive information of providers in your area code and a description of their services.

> **NOTE**
> All providers that service your area will be found by the area code listing.

Tips For Finding a Provider

If your area code is not listed
There are providers who have nationwide access. Some of the Web sites have information on these service providers.

If there is no local dial-in number
Look for service providers that are the closest to you or who have an 800–number dial-in access. Many providers are also listed on these Web sites.

Choosing a provider
Contact providers by phone, fax, or electronic mail. If you want to use Netscape or Explorer you will need to get a SLIP or a PPP account.

Ask about the following:

- Type of Internet accounts available.

- Price and hours of access. Ask how much it will cost per month for a SLIP or PPP account? How many hours of Internet access are included? An average price is $20 per month for 150 hours of graphical access.

- Technical support. Does the provider offer technical support? What are the hours (days, nights, weekends, holidays)? Is support free?

- Software. Do they provide the TCP/IP software? Is the software custom configured? Do they provide free copies of an e-mail program such as Eudora or a Web browser such as Netscape Navigator or Explorer? Good Internet providers will provide custom configured TCP/IP software and the essential Internet navigation and communication software.

NOTE

If you are using Windows 95 you will need to get information for configuring your Windows 95 TCP/IP software. At the time of this printing you cannot get TCP/IP software custom configured for Windows 95.

APPENDIX III
Using An Internet Navigational Suite

Internet front-end navigational suites are complete packages of tools that make it easier for you to connect to the Internet. In the past these suites provided separate software applications packaged together. The newer versions offer integrated software programs that are simple and save time. Every aspect of the Internet is easier, including your initial Internet set-up, Internet navigation, and downloading files using the File Transfer Protocol (FTP).

All of the front-end packages include the following:

- A configuration utility for establishing your Internet service
- E-mail software
- A graphical Web browser
- A newsgroup reader
- An FTP utility

The configuration utility assists you with dialing up a service provider and opening an account. The service providers listed in the software are usually limited to several large companies.

NOTE
- The cost for an Internet connection provided by the companies listed in front-end suites may be more expensive than the cost of using a local Internet provider.

- Integrated software packages may not allow you to use other e-mail programs or Web browsers.

Suggested Internet Navigational Suites

- EXPLORE Internet: (800) 863-4548
- IBM Internet Connection For Windows: (800) 354-3222
- Internet Anywhere: (519) 888-9910
- Internet Chameleon: (408) 973-7171
- Internet In A Box: (800) 557-9614 or (800) 777-9638
- Netscape Navigator Personal Edition: (415) 528-2555
- SuperHighway Access: (800) 929-3054

INDEX